Plant Cell Biology

Plant Cell Biology

Edited by
Mark Harper

Larsen & Keller
www.larsen-keller.com

Plant Cell Biology
Edited by Mark Harper
ISBN: 978-1-63549-648-2 (Hardback)

🖿 Larsen & Keller

Published by Larsen and Keller Education,
5 Penn Plaza,
19th Floor,
New York, NY 10001, USA

Cataloging-in-Publication Data

Plant cell biology / edited by Mark Harper.
     p. cm.
Includes bibliographical references and index.
ISBN 978-1-63549-648-2
1. Plants--Cytology. 2. Plant cells and tissues. 3. Plant cell biotechnology. 4. Cytology.
I. Harper, Mark.
QK725 .P53 2018
571.6--dc23

For more information regarding Larsen and Keller Education and its products, please visit the publisher's website www.larsen-keller.com

Table of Contents

Permissions

Index

Preface

Plant cells are types of eukaryotic cells. They have cell walls, plastids, cell plate, central vacuole, etc. They have various cell types, and tissue types. Plant cells can primarily be categorized into parenchyma cells and collenchyma cells. This book will provide the readers in-depth knowledge about the characteristics of plant cells. Some of the diverse topics covered in it address the varied branches that fall under this category. This textbook attempts to assist those with a goal of delving into the field of plant cell biology.

A foreword of all chapters of the book is provided below:

Chapter 1 - Plant cells are eukaryotic cells that have certain characteristics which are different from animal cells. The different types of cells are parenchyma cells, sclerenchyma cells and collenchyma cells. The chapter on plant cell offers an insightful focus, keeping in mind the complex subject matter; **Chapter 2** - The type of cell that is out in the outer epidermal layer of a plant is known as pavement cells. These cells are mainly simple in nature and have no specialized function. Palisade cell, guard cell, statocyte, ground tissue, megaspore mother cell, transfer cell, vascular cambium, sclereid, xylem and aerenchyma are the other types of plant cells and tissues explained in the chapter. The major categories of plant cells are dealt with great details in the chapter; **Chapter 3** - Plant cell wall is a multi-layered structure which surrounds some cells. The wall can be flexible and tough. The plant cell contains cellulose, pectin, lignin, plastid and plasmodesma. Cellulose is a polysaccharide that contain thousands of D-glucose units. This chapter is an overview of the subject matter incorporating all the major aspects of plant cell wall; **Chapter 4** - The plastid is a double membrane organelle which contain pigments used in the process of photosynthesis. Gerontoplast, chromoplast, leucoplast and proteinoplast are some topics discussed in the chapter. This chapter provides a plethora of interdisciplinary topics for better comprehension of plastids; **Chapter 5** - The essential aspects of plant cell biology are plasmolysis, turgor pressure, oleosin and stoma. Turgor pressure is the force in the cell which helps in pushing the plasma membrane against the cell wall. The aspects elucidated in this chapter are of vital importance, and provide a better understanding of plant cell biology.

At the end, I would like to thank all the people associated with this book devoting their precious time and providing their valuable contributions to this book. I would also like to express my gratitude to my fellow colleagues who encouraged me throughout the process.

Editor

An Introduction to Plant Cell

Plant cells are eukaryotic cells that have certain characteristics which are different from animal cells. The different types of cells are parenchyma cells, sclerenchyma cells and collenchyma cells. The chapter on plant cell offers an insightful focus, keeping in mind the complex subject matter.

Plant Cell

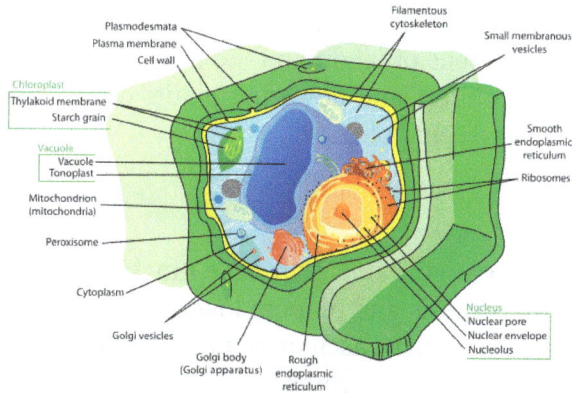

Plant cell structure

Plant cells are eukaryotic cells that differ in several key aspects from the cells of other eukaryotic organisms. Their distinctive features include:

- A large central vacuole, a water-filled volume enclosed by a membrane known as the *tonoplast* that maintains the cell's turgor, controls movement of molecules between the cytosol and sap, stores useful material and digests waste proteins and organelles.

- A cell wall composed of cellulose and hemicelluloses, pectin and in many cases lignin, is secreted by the protoplast on the outside of the cell membrane. This contrasts with the cell walls of fungi, which are made of chitin, and of bacteria, which are made of peptidoglycan. Cell walls perform many essential functions: they provide shape to form the tissue and organs of the plant, and play an important role in intercellular communication and plant-microbe interactions.

- Specialized cell-to-cell communication pathways known as plasmodesmata, pores in the primary cell wall through which the plasmalemma and endoplasmic reticulum of adjacent cells are continuous.

- Plastids, the most notable being the chloroplast, which contains chlorophyll, a green-colored pigment that absorbs sunlight, and allows the plant to make its own food in the process known as photosynthesis. Other types of plastids are the amyloplasts, specialized for starch storage, elaioplasts specialized for fat storage, and chromoplasts specialized for synthesis and storage of pigments. As in mitochondria, which have a genome encoding 37 genes, plastids have their own genomes of about 100–120 unique genes and, it is presumed, arose as prokaryoticendosymbionts living in the cells of an early eukaryotic ancestor of the land plants and algae.

- Cell division by construction of a phragmoplast as a template for building a cell plate late in cytokinesis is characteristic of land plants and a few groups of algae, notably the Charophytes and the Chlorophyte Order Trentepohliales.

- The motile, free-swimming sperm of bryophytes and pteridophytes, cycads and *Ginkgo* are the only cells of land plants to have flagella similar to those in animal cells, but the conifers and flowering plants do not have motile sperm and lack both flagella and centrioles.

Cell Types

- Parenchyma cells are living cells that have functions ranging from storage and support to photosynthesis and phloem loading (transfer cells). Apart from the xylem and phloem in their vascular bundles, leaves are composed mainly of parenchyma cells. Some parenchyma cells, as in the epidermis, are specialized for light penetration and focusing or regulation of gas exchange, but others are among the least specialized cells in plant tissue, and may remain totipotent, capable of dividing to produce new populations of undifferentiated cells, throughout their lives. Parenchyma cells have thin, permeable primary walls enabling the transport of small molecules between them, and their cytoplasm is responsible for a wide range of biochemical functions such as nectarsecretion, or the manufacture of secondary products that discourage herbivory. Parenchyma cells that contain many chloroplasts and are concerned primarily with photosynthesis are called chlorenchyma cells. Others, such as the majority of the parenchyma cells in potatotubers and the seedcotyledons of legumes, have a storage function.

- Collenchyma cells – collenchyma cells are alive at maturity and have only a primary wall. These cells mature from meristem derivatives that initially resemble parenchyma, but differences quickly become apparent. Plastids do not develop, and the secretory apparatus (ER and Golgi) proliferates to secrete additional primary wall. The wall is most commonly thickest at the corners, where three or more cells come in contact, and thinnest where only two cells come in contact, though other arrangements of the wall thickening are possible.

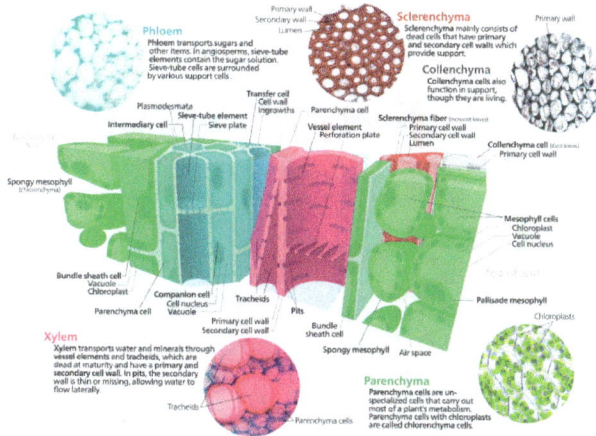

Cross section of a leaf showing various plant cell types

Pectin and hemicellulose are the dominant constituents of collenchyma cell walls of dicotyledonangiosperms, which may contain as little as 20% of cellulose in *Petasites*. Collenchyma cells are typically quite elongated, and may divide transversely to give a septate appearance. The role of this cell type is to support the plant in axes still growing in length, and to confer flexibility and tensile strength on tissues. The primary wall lacks lignin that would make it tough and rigid, so this cell type provides what could be called plastic support – support that can hold a young stem or petiole into the air, but in cells that can be stretched as the cells around them elongate. Stretchable support (without elastic snap-back) is a good way to describe what collenchyma does. Parts of the strings in celery are collenchyma.

- Sclerenchyma cells – Sclerenchyma cells (from the Greek skleros, *hard*) are hard and tough cells with a function in mechanical support. They are of two broad types – sclereids or stone cells and fibres. The cells develop an extensive secondary cell wall that is laid down on the inside of the primary cell wall. The secondary wall is impregnated with lignin, making it hard and impermeable to water. Thus, these cells cannot survive for long' as they cannot exchange sufficient material to maintain active metabolism. Sclerenchyma cells are typically dead at functional maturity, and the cytoplasm is missing, leaving an empty central cavity.

Functions for sclereid cells (hard cells that give leaves or fruits a gritty texture) include discouraging herbivory, by damaging digestive passages in small insect larval stages, and physical protection (a solid tissue of hard sclereid cells form the pit wall in a peach and many other fruits). Functions of fibres include provision of load-bearing support and tensile strength to the leaves and stems of herbaceous plants. Sclerenchyma fibres are not involved in conduction, either of water and nutrients (as in the xylem) or of carbon compounds (as in the phloem), but it is likely that they may have evolved as modifications of xylem and phloem initials in early land plants.

Tissue Types

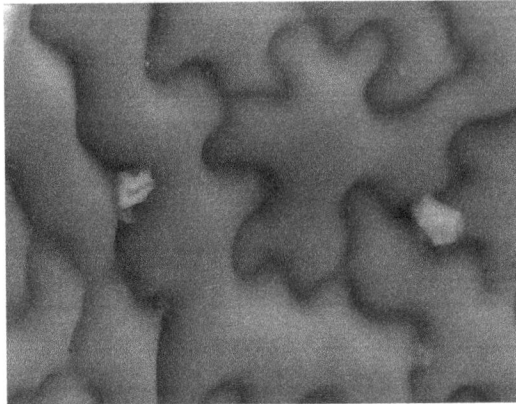

Cells of *Arabidopsis thaliana* epidermis

The major classes of cells differentiate from undifferentiated meristematic cells (analogous to the stem cells of animals) to form the tissue structures of roots, stems, leaves, flowers, and reproductive structures.

Xylem cells are elongated cells with lignified secondary thickening of the cell walls. Xylem cells are specialised for conduction of water, and first appeared in plants during their transition to land in the Silurian period more than 425 million years ago. The possession of xylem defines the vascular plants or Tracheophytes. Xylem tracheids are pointed, elongated xylem cells, the simplest of which have continuous primary cell walls and lignified secondary wall thickenings in the form of rings, hoops, or reticulate networks. More complex tracheids with valve-like perforations called bordered pits characterise the gymnosperms. The ferns and other pteridophytes and the gymnosperms have only xylem tracheids, while the angiosperms also have xylem vessels. Vessel members are hollow xylem cells without end walls that are aligned end-to-end so as to form long continuous tubes. The bryophytes lack true xylem cells, but their sporophytes have a water-conducting tissue known as the hydrome that is composed of elongated cells of simpler construction.

Phloem is a specialised tissue for food transport in higher plants. Phloem cells mainly transport sucrose along pressure gradients generated by osmosis. This phenomenon is called translocation. Phloem consists of two cell types, the sieve tubes and the intimately associated companion cells. The sieve tube elements lack nuclei and ribosomes, and their metabolism and functions are regulated by the adjacent nucleate companion cells. Sieve tubes are joined end-to-end with perforate end-plates between known as *sieve plates*, which allow transport of photosynthate between the sieve elements. The companion cells, connected to the sieve tubes via plasmodesmata, are responsible for loading the phloem with sugars. The bryophytes lack phloem, but moss sporophytes have a simpler tissue with analogous function known as the leptome.

Plant epidermal cells are specialised parenchyma cells covering the external surfaces of leaves, stems and roots. The epidermal cells of aerial organs arise from the superficial layer of cells known as the *tunica* (L1 and L2 layers) that covers the plant shoot apex, whereas the cortex and vascular tissues arise from innermost layer of the shoot apex known as the *corpus* (L3 layer). The epidermis of roots originates from the layer of cells immediately beneath the root cap.

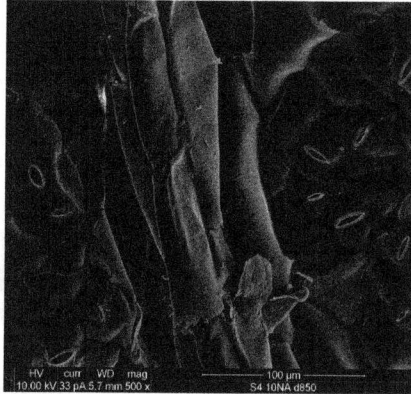

This is an electron micrograph of the epidermal cells of a Brassica chinensis leaf.
The stomates are also visible.

The epidermis of all aerial organs, but not roots, is covered with a cuticle made of the polyestercutin and/or the hydrocarbon polymer cutan with a superficial layer of epicuticular waxes. The epidermal cells of the primary shoot are thought to be the only plant cells with the biochemical capacity to synthesize cutin. Several cell types may be present in the epidermis. Notable among these are the stomatal guard cells, glandular and clothing hairs or trichomes, and the root hairs of primary roots. In the shoot epidermis of most plants, only the guard cells have chloroplasts. Chloroplasts contain the green pigment chlorophyll which is needed for photosynthesis.

Organelles

• Cell membrane	• Chloroplast	• Nucleus
• Cell wall	• Leucoplast	• Chromatin
• Nuclear membrane	• Chromoplast	• Cytoskeleton
• Vacuole	• Golgi Bodies	• Nucleolus
• Plastid	• Cytoplasm	• Mitochondrion

Phragmoplast

The phragmoplast is a plant cell specific structure that forms during late cytokinesis. It serves as a scaffold for cell plate assembly and subsequent formation of a new cell wall separating the two daughter cells.

The phragmoplast is a complex assembly of microtubules (MTs), microfilaments (MFs), and endoplasmic reticulum (ER) elements, that assemble in two opposing sets perpendicular to the plane of the future cell plate during anaphase and telophase. It is initially barrel-shaped and forms from the mitotic spindle between the two daughter nuclei while nuclear envelopes reassemble around them. The cell plate initially forms as a disc between the two halves of the phragmoplast structure. While new cell plate material is added to the edges of the growing plate, the phragmoplast microtubules disappear in the center and regenerate at the edges of the growing cell plate. The two structures grow outwards until they reach the outer wall of the dividing cell. If a phragmosome was present in the cell, the phragmoplast and cell plate will grow through the space occupied by the phragmosome. They will reach the parent cell wall exactly at the position formerly occupied by the preprophase band.

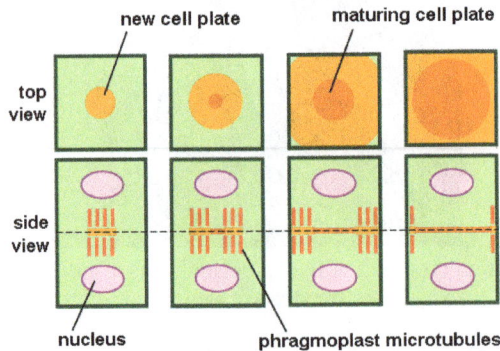

Phragmoplast and cell plate formation in a plant cell during cytokinesis. Left side: Phragmoplast forms and cell plate starts to assemble in the center of the cell. Towards the right: Phragmoplast enlarges in a donut-shape towards the outside of the cell, leaving behind mature cell plate in the center. The cell plate will transform into the new cell wall once cytokinesis is complete.

The microtubules and actin filaments within the phragmoplast serve to guide vesicles with cell wall material to the growing cell plate. Actin filaments are also possibly involved in guiding the phragmoplast to the site of the former preprophase band location at the parent cell wall. While the cell plate is growing, segments of smooth endoplasmic reticulum are trapped within it, later forming the plasmodesmata connecting the two daughter cells.

The phragmoplast can only be observed in Embryophytes, that is the bryophytes and vascular plants, and a few advanced green algae, specifically *Coleochaete* in the Division Charophyta. Some algae use another type of microtubule array, a phycoplast, during cytokinesis.

Tannosome

Tannosomes are organelles found in plant cells of vascular plants.

Formation and Functions

Tannosomes are formed when the chloroplast membrane forms pockets filled with tannin. Slowly, the pockets break off as tiny vacuoles that carry tannin and to the large vacuole filled with acidic fluid. Tannins are then released into the vacuole and stored inside as tannin accretions.

They are responsible for synthesizing and producing condensed tannins and poly-phenols. Tannosomes condense tannins in chlorophyllous organs, providing defenses against herbivores and pathogens, and protection against UV radiation.

Discovery

Tannosomes were discovered in September 2013 by French and Hungarian researchers.

The discovery of tannosomes showed how to get enough tannins to change the flavour of wine, tea, chocolate, and other food or snacks.

Epidermis (Botany)

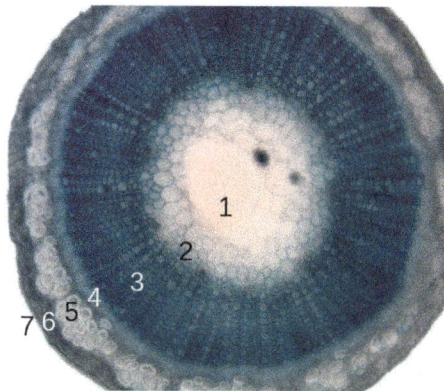

Cross-section of a flax plant stem:
1. pith 2. protoxylem 3. xylem 4. phloem 5. sclerenchyma (bast fibre) 6. cortex 7. epidermis

The epidermis is a single layer of cells that covers the leaves, flowers, roots and stems of plants. It forms a boundary between the plant and the external environment. The epidermis serves several functions: it protects against water loss, regulates gas exchange, secretes metabolic compounds, and (especially in roots) absorbs water and mineral nutrients. The epidermis of most leaves shows dorsoventral anatomy: the upper (adaxial) and lower (abaxial) surfaces have somewhat different construction and may serve different functions. Woody stems and some other stem structures produce a secondary covering called the periderm that replaces the epidermis as the protective covering.

Description

The epidermis is the outermost cell layer of the primary plant body. In some older works the cells of the leaf epidermis have been regarded as specialised parenchyma cells, but the established modern preference has long been to classify the epidermis as dermal tissue, whereas parenchyma is classified as ground tissue. The epidermis is the main component of the dermal tissue system of leaves, and also stems, roots, flowers, fruits, and seeds; it is usually transparent (epidermal cells have fewer chloroplasts or lack them completely, except for the guard cells.)

The cells of the epidermis are structurally and functionally variable. Most plants have an epidermis that is a single cell layer thick. Some plants like *Ficus elastica* and *Peperomia*, which have periclinal cellular division within the protoderm of the leaves, have an epidermis with multiple cell layers. Epidermal cells are tightly linked to each other and provide mechanical strength and protection to the plant. The walls of the epidermal cells of the above ground parts of plants contain cutin, and are covered with a cuticle. The cuticle reduces water loss to the atmosphere, it is sometimes covered with wax in smooth sheets, granules, plates, tubes or filaments. The wax layers give some plants a whitish or bluish surface color. Surface wax acts as a moisture barrier and protects the plant from intense sunlight and wind. The underside of many leaves have a thinner cuticle than the top side, and leaves of plants from dry climates often have thickened cuticles to conserve water by reducing transpiration.

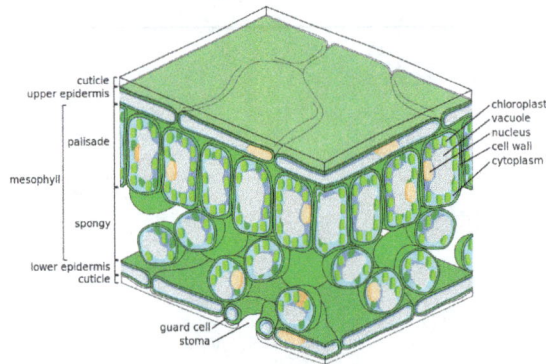

The epidermal tissue includes several differentiated cell types: epidermal cells, guard cells, subsidiary cells, and epidermal hairs (trichomes). The epidermal cells are the most numerous, largest, and least specialized. These are typically more elongated in the leaves of monocots than in those of dicots.

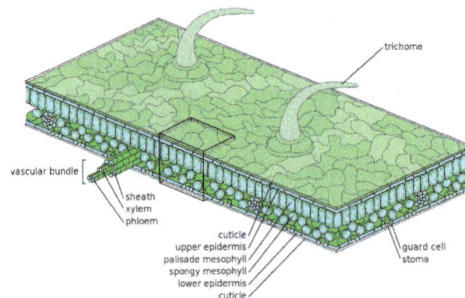

Trichomes or hairs grow out from the epidermis in many species. In root epidermis, epidermal hairs, termed root hairs are common and are specialized for absorption of water and mineral nutrients.

In plants with secondary growth, the epidermis of roots and stems is usually replaced by a periderm through the action of a cork cambium.

Guard Cells

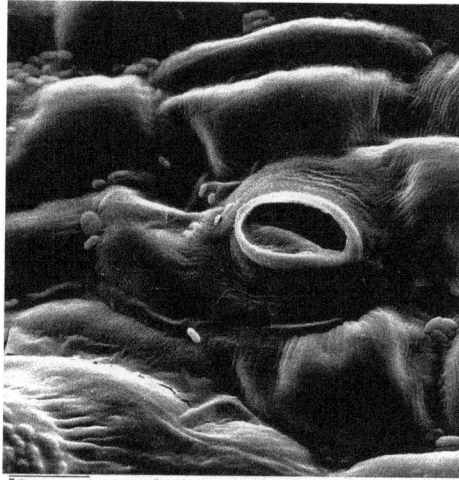

Stoma in a tomato leaf (microscope image)

The leaf and stem epidermis is covered with pores called stomata (sing., stoma), part of a stoma complex consisting of a pore surrounded on each side by chloroplast-containing guard cells, and two to four subsidiary cells that lack chloroplasts. The stomata complex regulates the exchange of gases and water vapor between the outside air and the interior of the leaf. Typically, the stomata are more numerous over the abaxial (lower) epidermis of the leaf than the (adaxial) upper epidermis. An exception is floating leaves where most or all stomata are on the upper surface. Vertical leaves, such as those of many grasses, often have roughly equal numbers of stomata on both surfaces. The stoma is bounded by two guard cells. The guard cells differ from the epidermal cells in the following aspects:

- The guard cells are bean-shaped in surface view, while the epidermal cells are irregular in shape.

- The guard cells contain chloroplasts, so they can manufacture food by photosynthesis (The epidermal cells do not contain chloroplasts).

- Guard Cells are the only epidermal cells that can make sugar. According to one theory, in sunlight the concentration of potassium ions (K+) increases in the guard cells. This, together with the sugars formed, lowers the water potential in the guard cells. As a result, water from other cells enter the guard cells by osmosis so they swell and become turgid. Because the guard cells have a thicker cellulose wall on one side of the cell, i.e. the side around the stomatal pore, the swollen guard cells become curved and pull the stomata open.

At night, the sugar is used up and water leaves the guard cells, so they become flaccid and the stomatal pore closes. In this way, they reduce the amount of water vapour escaping from the leaf.

Cell Differentiation in the Epidermis

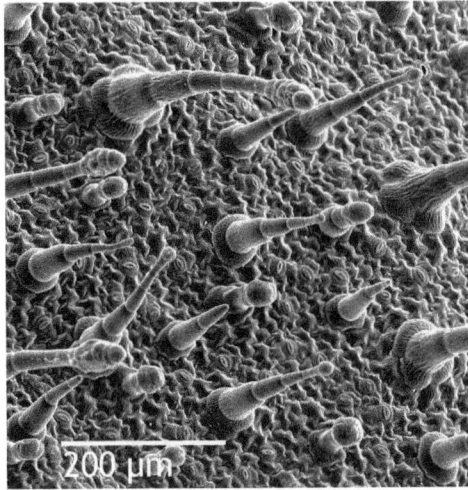

Scanning electron microscope image of *Nicotiana alata* leaf's epidermis, showing trichomes (hair-like appendages) and stomata (eye-shaped slits, visible at full resolution).

The plant epidermis consists of three main cell types: pavement cells, guard cells and their subsidiary cells that surround the stomata and trichomes, otherwise known as leaf hairs. The epidermis of petals also form a variation of trichomes called conical cells.

Trichomes develop at a distinct phase during leaf development, under the control of two major trichome specification genes: *TTG* and *GL1*. The process may be controlled by the plant hormonesgibberellins, and even if not completely controlled, gibberellins certainly have an effect on the development of the leaf hairs. *GL1* causes endoreplication, the replication of DNA without subsequent cell division as well as cell expansion. *GL1* turns on the expression of a second gene for trichome formation, *GL2*, which controls the final stages of trichome formation causing the cellular outgrowth.

Arabidopsis thaliana uses the products of inhibitory genes to control the patterning of trichomes, such as *TTG* and *TRY*. The products of these genes will diffuse into the lateral cells, preventing them from forming trichomes and in the case of *TRY* promoting the formation of pavement cells.

Expression of the gene *MIXTA*, or its analogue in other species, later in the process of cellular differentiation will cause the formation of conical cells over trichomes. *MIXTA* is a transcription factor.

Stomatal patterning is a much more controlled process, as the stoma effect the plants water retention and respiration capabilities. As a consequence of these important functions, differentiation of cells to form stomata is also subject to environmental conditions to a much greater degree than other epidermal cell types.

Stomata are pores in the plant epidermis that are surrounded by two guard cells, which control the opening and closing of the aperture. These guard cells are in turn surrounded by subsidiary cells which provide a supporting role for the guard cells.

Stomata begin as stomatal meristemoids. The process varies between dicots and monocots. Spacing is thought to be essentially random in dicots though mutants do show it is under some form of genetic control, but it is more controlled in monocots, where stomata arise from specific asymmetric divisions of protoderm cells. The smaller of the two cells produced becomes the guard mother cells. Adjacent epidermal cells will also divide asymmetrically to form the subsidiary cells.

Because stomata play such an important role in the plants survival, collecting information on their differentiation is difficult by the traditional means of genetic manipulation, as stomatal mutants tend to be unable to survive. Thus the control of the process is not well understood. Some genes have been identified. *TMM* is thought to control the timing of stomatal initiation specification and *FLP* is thought to be involved in preventing further division of the guard cells once they are formed.

Environmental conditions affect the development of stomata, in particular their density on the leaf surface. It is thought that plant hormones, such as ethylene and cytokines, control the stomatal developmental response to the environmental conditions. Accumulation of these hormones appears to cause increased stomatal density such as when the plants are kept in closed environments.

References

- P.H. Raven, R.F. Evert, S.E. Eichhorn (2005): Biology of Plants, 7th Edition, W.H. Freeman and Company Publishers, New York, ISBN 0-7167-1007-2

- Brillouet, J. -M.; Romieu, C.; Schoefs, B.; Solymosi, K.; Cheynier, V.; Fulcrand, H.; Verdeil, J. -L.; Conejero, G. (2013). "The tannosome is an organelle forming condensed tannins in the chlorophyllous organs of Tracheophyta". Annals of Botany. 112 (6): 1003. PMC 3783233. PMID 24026439. doi:10.1093/aob/mct168

- Evert, Ray F; Eichhorn, Susan E. Esau's Plant Anatomy: Meristems, Cells, and Tissues of the Plant Body: Their Structure, Function, and Development. Publisher: Wiley-Liss 2006. ISBN 978-0471738435

- Kerry Grens (2013-09-23). "New Organelle: The Tannosome". The Scientist Magazine®. Retrieved 2014-08-22

- Raven, Peter H.; Evert, Ray F.; Curtis, Helena (1981), Biology of plants, New York, N.Y.: Worth Publishers, pp. 427–28, ISBN 0-87901-132-7, OCLC 222047616

Types of Plant Cells and Tissues

The type of cell that is out in the outer epidermal layer of a plant is known as pavement cells. These cells are mainly simple in nature and have no specialized function. Palisade cell, guard cell, statocyte, ground tissue, megaspore mother cell, transfer cell, vascular cambium, sclereid, xylem and aerenchyma are the other types of plant cells and tissues explained in the chapter. The major categories of plant cells are dealt with great details in the chapter.

Pavement Cells

Pavement cells are a cell type found in the outmost epidermal layer of plants. They are simple cells with no specialized function. Together, the main purpose of these cells is to form a protective layer for the more specialized cells below. This layer helps decrease water loss, maintain an internal temperature, keep the inner cells in place, and resist the intrusion of any outside material. They also separate stomata apart from each other as stomata have at least one pavement cell between each other.

They do not have a regular shape. Rather, their irregular shapes help them to interlock with each other like puzzle pieces to form a sturdy layer. This irregular shape that each individual cell takes on can be influenced by the cytoskeleton and specific proteins. As the leaf grows, the pavement cells will also grow, divide, and synthesize new vacuoles, plasma membrane parts, and cell wall components. A thick external cell wall influences the direction of growth by impeding expansion towards the outside of the cell and instead promote expansion parallel to the epidermis layer.

Palisade Cell

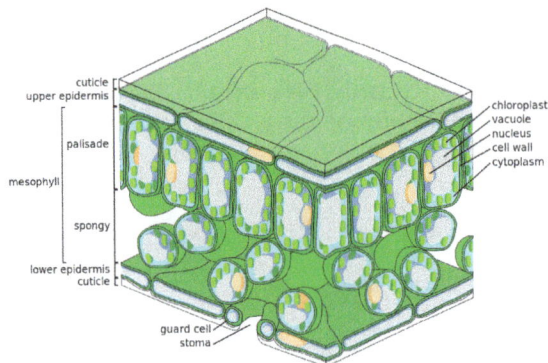

Diagram of the internal structure of a leaf

Palisade cells are plant cells located within the mesophyll in leaves, right below the upper epidermis and cuticle. They are vertically elongated, a different shape from the spongy mesophyll cells beneath them in the leaf. Their chloroplasts absorb a major portion of the light energy used by the leaf. Palisade cells occur in dicotyledonous plants, and also in the net-veined monocots, the Araceae and Dioscoreaceae.

Structure

Palisade cells contain the largest number of chloroplasts per cell, which makes them the primary site of photosynthesis in the leaves of those plants that contain them, converting the energy in light to the chemical energy of carbohydrates

Beneath the palisade mesophyll are the spongy mesophyll cells, which also perform photosynthesis. They are irregularly shaped cells that have many intercellular spaces that allow the passage of gases, such as the carbon dioxide needed for photosynthesis.

Palisade cells are chlorenchyma cells, i.e., parenchyma cells containing chloroplasts.

Guard Cell

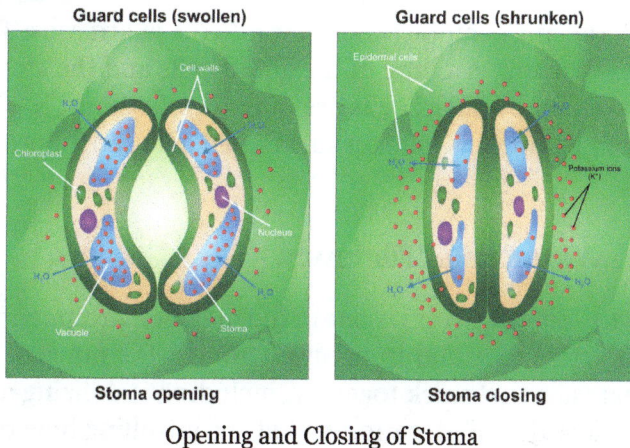

Guard cells (swollen)   Guard cells (shrunken)

Stoma opening   Stoma closing

Opening and Closing of Stoma

Guard cells are specialized cells in the epidermis of leaves, stems and other organs that are used to control gas exchange. They are produced in pairs with a gap between them that forms a stomatal pore. The stomatal pores are largest when water is freely available and the guard cells turgid, and closed when water availability is critically low and the guard cells become flaccid. Photosynthesis depends on the diffusion of carbon dioxide (CO_2) from the air through the stomata into the mesophyll tissues. Oxygen (O_2), produced as a byproduct of photosynthesis, exits the plant via the stomata. When the stomata are open, water is lost by evaporation and must be replaced via the transpiration stream, with water taken up by the roots. Plants must balance the amount of

CO_2 absorbed from the air with the water loss through the stomatal pores, and this is achieved by both active and passive control of guard cell turgor and stomatal pore size.

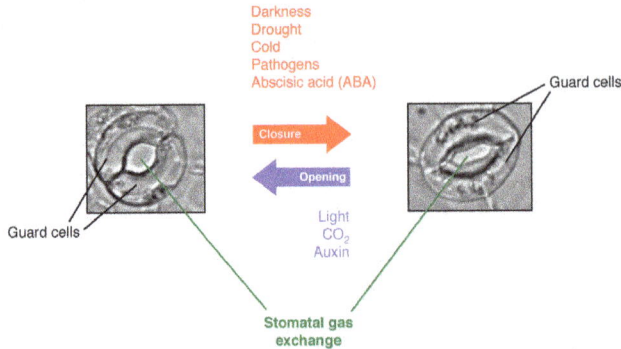

A stomatal pore in the surface (epidermis) of a leaf as viewed through a microscope. The central stomatal pore is formed by a pair of guard cells. The stomatal pore can either open (left) or close (right) depending on the environmental conditions.

Guard Cell Function

Opening and closure of the stomatal pore is mediated by changes in the turgor pressure of the two guard cells.The turgor pressure of guard cells is controlled by movements of large quantities of ions and sugars into and out of the guard cells. When guard cells take up these solutes, the water potential (Ψ) inside the cells decreases (creating a hypotonic solution), causing osmotic water flow into the guard cells. This leads to a turgor pressure increase causing swelling of the guard cells and the stomatal pores open. The ions that are taken up by guard cells are mainly potassium (K^+) ions and chloride (Cl^-) ions. In addition guard cells take up sugars that also contribute to opening of the stomatal pores.

Water Loss and Water use Efficiency

Water stress (drought and salt stress) is one of the major environmental problems causing severe losses in agriculture and in nature. Drought tolerance of plants is mediated by several mechanisms that work together, including stabilizing and protecting the plant from damage caused by desiccation and also controlling how much water plants lose through the stomatal pores during drought. A plant hormone, abscisic acid (ABA), is produced in response to drought. A major type of ABA receptor has been identified. Future research is needed to test if these receptors can be used to engineer drought tolerance in plants. The plant hormone ABA causes the stomatal pores to close in response to drought, which reduces plant water loss via transpiration to the atmosphere and allows plants to avoid or slow down water loss during droughts. The use of drought tolerant crop plants would lead to a reduction in crop losses during droughts. Since guard cells control water loss of plants, the investigation on how stomatal opening and closure are regulated could lead to the development of plants with improved avoidance or slowing of desiccation and better water use efficiency.

Ion Uptake and Release

Ion uptake into guard cells causes stomatal opening: The opening of gas exchange pores requires the uptake of potassium ions into guard cells. Potassium channels and pumps have been identified and shown to function in the uptake of ions and opening of stomatal apertures. Ion release from guard cells causes stomatal pore closing: Other ion channels have been identified that mediate release of ions from guard cells, which results in osmotic water efflux from guard cells due to osmosis, shrinking of the guard cells, and closing of stomatal pores. Specialized potassium efflux channels participate in mediating release of potassium from guard cells. Anion channels were identified as important controllers of stomatal closing. Anion channels have several major functions in controlling stomatal closing: (a) They allow release of anions, such as chloride and malate from guard cells, which is needed for stomatal closing. (b) Anion channels are activated by signals that cause stomatal closing, for example by intracellular calcium and ABA. The resulting release of negatively charged anions from guard cells results in an electrical shift of the membrane to more positive voltages (depolarization) at the intracellular surface of the guard cell plasma membrane. This electrical depolarization of guard cells leads to activation of the outward potassium channels and the release of potassium through these channels. At least two major types of anion channels have been characterized in the plasma membrane: S-type anion channels and R-type anion channels.

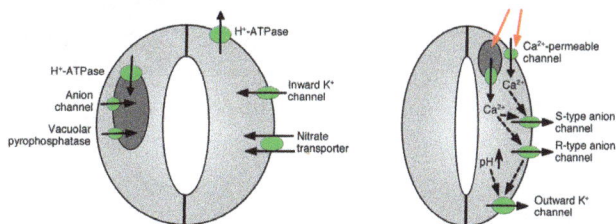

Ion channels and pumps regulating stomatal opening and closure

Vacuolar Ion Transport

Vacuoles are large intracellular storage organelles in plants cells. In addition to the ion channels in the plasma membrane, vacuolar ion channels have important functions in regulation of stomatal opening and closure because vacuoles can occupy up to 90% of guard cell's volume. Therefore, a majority of ions are released from vacuoles when stomata are closed. Vascuolar K^+ (VK) channels and fast vacuolar channels can mediate K^+ release from vacuoles. Vacuolar K^+ (VK) channels are activated by elevation in the intracellular calcium concentration. Another type of calcium-activated channel, is the slow vacuolar (SV) channel. SV channels have been shown to function as cation channels that are permeable to Ca^{2+} ions, but their exact functions are not yet known in plants.

Signal Transduction

Guard cells perceive and process environmental and endogenous stimuli such as light, humidity, CO_2 concentration, temperature, drought, and plant hormones to trigger cellular responses resulting in stomatal opening or closure. These signal transduction pathways determine for example how quickly a plant will lose water during a drought period. Guard cells have become a model for single cell signaling. Using *Arabidopsis thaliana*, the investigation of signal processing in single guard cells has become open to the power of genetics. Cytosolic and nuclear proteins and chemical messengers that function in stomatal movements have been identified that mediate the transduction of environmental signals thus controlling CO_2 intake into plants and plant water loss. Research on guard cell signal transduction mechanisms is producing an understanding of how plants can improve their response to drought stress by reducing plant water loss. Guard cells also provide an excellent model for basic studies on how a cell integrates numerous kinds of input signals to produce a response (stomatal opening or closing). These responses require coordination of numerous cell biological processes in guard cells, including signal reception, ion channel and pump regulation, membrane trafficking, transcription, cytoskeletal rearrangements and more. A challenge for future research is to assign the functions of some of the identified proteins to these diverse cell biological processes.

Development

During the development of plant leaves, the specialized guard cells differentiate from "guard mother cells". The density of the stomatal pores in leaves is regulated by environmental signals, including increasing atmospheric CO_2 concentration, which reduces the density of stomatal pores in the surface of leaves in many plant species by presently unknown mechanisms. The genetics of stomatal development can be directly studied by imaging of the leaf epidermis using a microscope. Several major control proteins that function in a pathway mediating the development of guard cells and the stomatal pores have been identified.

Megaspore Mother Cell

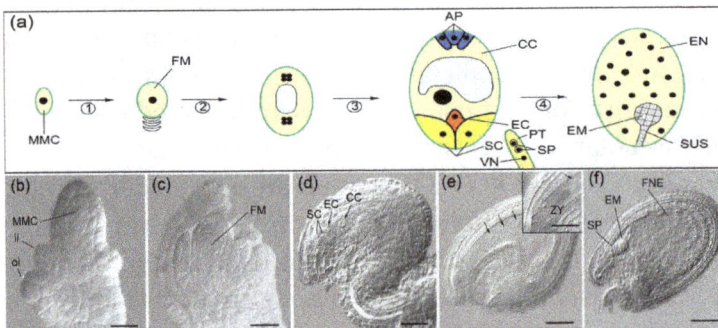

Development of the megagametophyte and fertilization in *Arabidopsis thaliana*.
MMC - megaspore mother cell; FM - functional megaspore; AP - antipodal cells;
CC - central cell; EC - egg cell; SC - synergid cell; PT - pollen tube; SP - sperm;
VN - vegetative nucleus; EM - embryo; EN - endosperm; SUS - suspensor.

A megaspore mother cell, or megasporocyte, is a diploid cell in plants in which meiosis will occur, resulting in the production of four haploidmegaspores. At least one of the spores develop into haploid female gametophytes (megagametophytes). The megaspore mother cell arises within the megasporangium tissue.

In flowering plants the megasporangium is also called the nucellus, and the female gametophyte is sometimes called the embryo sac.

Developmental Processes

Two distinct processes are involved in producing the megagametophyte from the megaspore mother cell:

- Megasporogenesis, formation of the megaspores in the megasporangium (nucellus) by meiosis

- Megagametogenesis; development of the megaspore(s) into the megagametophyte(s) which contains the gametes.

In gymnosperms and most flowering plants, only one of the four megaspores is functional at maturity, and the other three soon degenerate. The megaspore that remains divides mitotically and develops into the gametophyte, which eventually produces one egg cell. In the most common type of megagametophyte development in flowering plants (the *Polygonum* type), three mitotic divisions are involved in producing the gametophyte, which has seven cells, one of which (the central cell) has two nuclei that later merge to make a diploid nucleus.

In flowering plants, double fertilization occurs, which involves two sperm fertilizing the two gametes inside the megagametophyte (the egg cell and the central cell) to produce the embryo and the endosperm.

Statocyte

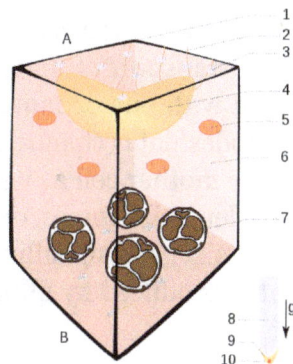

A root statocyte in a vertical position: (1) Cell Wall (2) Endoplasmic Reticulum (3) Plasmodesma (4) Nucleus (5) Mitochondrion (6) Cytoplasm (7) Statolith (8) Root (9) Root cap (10) Statocyte.

Statocytes are cells thought to be involved in gravitropic perception in plants, located in the cap tissue of the roots. They contain statoliths – starch-filled amyloplastic organelles – which sediment at the lowest part of the cells and initiate differential growth patterns, bending the root towards the vertical axis.

Sieve Tube Element

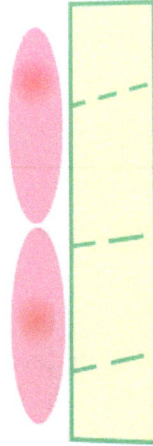

Sieve Tube and Companion Cell: Solid green-sieve tube; dashed green-sieve tube plates; light pink-companion cell; dark pink-nucleus; yellow-dissolved nutrients.

In plant anatomy, sieve tube elements, also called sieve tube members, are a specialised type of elongated cell in the phloem tissue of flowering plants. The ends of these cells are connected with other sieve tube members, and together they constitute the sieve tube. The main function of the sieve tube is transport of carbohydrates, primarily sucrose, in the plant (e.g., from the leaves to the fruits and roots). Unlike the water-conducting xylemvessel elements that are dead when mature, sieve elements are living cells. They are unique in lacking a nucleus at maturity.

At the interface between two sieve tube members in angiosperms are sieve plates, pores in the plant cell walls that facilitate transport of materials between them. Each sieve tube element is normally associated with one or more nucleated companion cells, to which they are connected by plasmodesmata (channels between the cells). Each companion cell is derived from the same mother cell as its associated sieve tube member. Sieve tube members have no cell nucleus, ribosomes, or vacuoles. Thus, they depend on companion cells to provide proteins, ATP, and signalling molecules. In leaves, companion cells help move the sugar that is produced by photosynthesis from the mesophyll tissue into the sieve tube elements.

Sieve cells are long, slender, conducting cells of the phloem that do not form a constituent element of a sieve tube, but which are provided with relatively unspecialized sieve areas, especially in the tapering ends of the cells that overlap those of other sieve cells. Sieve cells are typically associated with gymnosperms, because angiosperms have the more derived sieve tube members and companion cells in their phloem. They have a narrower diameter and are more elongated compared to sieve tube members. Sieve cells are associated with albuminous cells (also called Strasburger cells), which lack starch, thus making it possible to differentiate them from phloem parenchyma.

The forest botanist Theodor Hartig was the first to discover and name these cells as *Siebfasern* (sieve fibres) and *Siebröhren* (sieve tubes) in 1837.

Ground Tissue

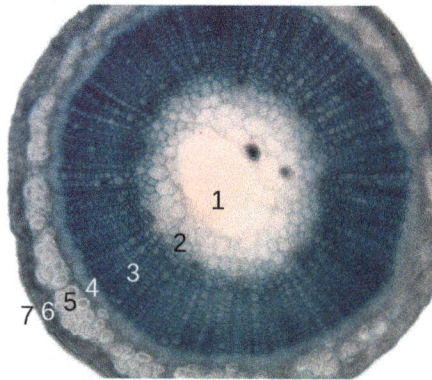

Cross-section of a flax plant stem:
1. pith 2. protoxylem 3. xylem 4. phloem 5. sclerenchyma (bast fibre) 6. cortex 7. epidermis

The ground tissue of plants includes all tissues that are neither dermal nor vascular. It can be divided into three classes based on the nature of the cell walls. Parenchyma cells have thin primary walls and usually remain alive after they become mature. Parenchyma forms the "filler" tissue in the soft parts of plants. Collenchyma cells have thin primary walls with some areas of secondary thickening. Collenchyma provides extra structural support, particularly in regions of new growth. Sclerenchyma cells have thick lignified secondary walls and often die when mature. Sclerenchyma provides the main structural support to a plant.

Parenchyma

Parenchyma is a versatile ground tissue that generally constitutes the "filler" tissue in soft parts of plants. It forms, among other things, the cortex and pith of stems, the cortex of roots, the mesophyll of leaves, the pulp of fruits, and the endosperm of seeds. Parenchyma cells are living cells and may remain meristematic at maturity—meaning that they are capable of cell division if stimulated. They have thin but flexible cellulosecell walls, and are generally polyhedral when close-packed, but can be roughly spherical when isolated from their neighbours. They have large central vacuoles, which allow the cells to store and regulate ions, waste products, and water. Tissue specialised for food storage is commonly formed of parenchyma cells.

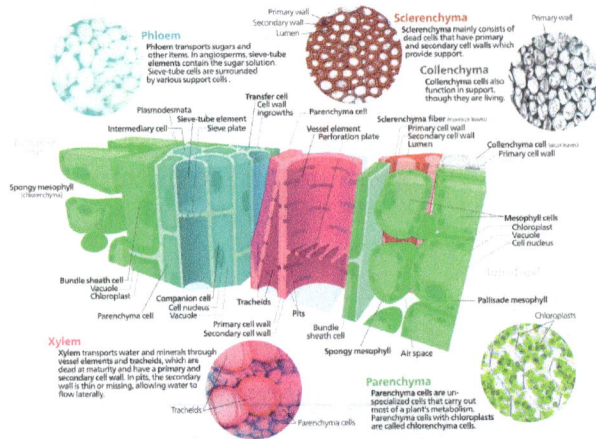

Cross section of a leaf showing various ground tissue types

Parenchyma cells have a variety of functions:

- In leaves, they form the mesophyll and are responsible for photosynthesis and the exchange of gases, parenchyma cells in the mesophyll of leaves are specialised parenchyma cells called chlorenchyma cells (parenchyma cells with chloroplasts).

- Storage of starch, protein, fats, oils and water in roots, tubers (e.g. potatoes), seed endosperm (e.g. cereals) and cotyledons (e.g. pulses and peanuts).

- Secretion (e.g. the parenchyma cells lining the inside of resin ducts).

- Wound repair and the potential for renewed meristematic activity.

- Other specialised functions such as aeration (aerenchyma) provides buoyancy and helps aquatic plants in floating.

- Chlorenchyma cells carry out photosynthesis and manufacture food.

The shape of parenchyma cells varies with their function. In the spongy mesophyll of a leaf, parenchyma cells range from near-spherical and loosely arranged with large intercellular spaces, to branched or stellate, mutually interconnected with their neighbours at the ends of their arms to form a three-dimensional network, like in the red kidney bean *Phaseolus vulgaris* and other mesophytes. These cells, along with the epidermalguard cells of the stoma, form a system of air spaces and chambers that regulate the exchange of gases. In some works the cells of the leaf epidermis are regarded as specialised parenchymal cells, but the modern preference has long been to classify the epidermis as plant dermal tissue, and parenchyma as ground tissue. shapes of parenchyma 1= polyhedral[these cells are most poly hedral shape] 2=stellate (found in stem of plants and have well developed air spaces between them)

3=elongated(are found in pallisade tissue of leaf) 4=lobed (are found inspongy and pallisade mesophyyll tissue of some plants)

Collenchyma

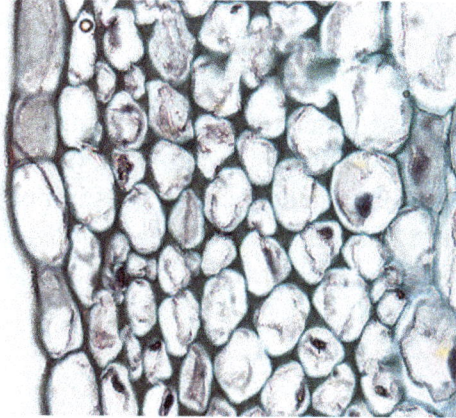

Cross section of collenchyma cells

The first use of "collenchyma" was by Link (1837) who used it to describe the sticky substance on *Bletia* (Orchidaceae) pollen. Complaining about Link's excessive nomenclature, Schleiden (1839) stated mockingly that the term "collenchyma" could have more easily been used to describe elongated sub-epidermal cells with unevenly thickened cell walls.

Collenchyma tissue is composed of elongated cells with irregularly thickened cell wall|lucky abhiram. They provide structural support, particularly in growing shoots and leaves. Collenchyma tissue makes up things such as the resilient strands in stalks of sai purnesh created in 1823.

Collenchyma cells are usually living, and have only a thick primary cell wall made up of cellulose and pectin. Cell wall thickness is strongly affected by mechanical stress upon the plant. The walls of collenchyma in shaken plants (to mimic the effects of wind etc.), may be 40–100% thicker than those not shaken.

There are four main types of collenchyma:

- Angular collenchyma (thickened at intercellular contact points)
- Tangential collenchyma (cells arranged into ordered rows and thickened at the tangential face of the cell wall)
- Annular collenchyma (uniformly thickened cell walls)
- Lacunar collenchyma (collenchyma with intercellular spaces)

Collenchyma cells are most often found adjacent to outer growing tissues such as the vascular cambium and are known for increasing structural support and integrity.

Sclerenchyma

Sclerenchyma is the supporting tissue in plants. Two types of sclerenchyma cells exist: fibres and sclereids. Their cell walls consist of cellulose, hemicellulose and lignin. Sclerenchyma cells are the principal supporting cells in plant tissues that have ceased elongation. Sclerenchyma fibres are of great economic importance, since they constitute the source material for many fabrics (e.g. flax, hemp, jute, and ramie).

Unlike the collenchyma, mature sclerenchyma is composed of dead cells with extremely thick cell walls (secondary walls) that make up to 90% of the whole cell volume. It is the hard, thick walls that make sclerenchyma cells important strengthening and supporting elements in plant parts that have ceased elongation. The difference between fibres and sclereids is not always clear: transitions do exist, sometimes even within the same plant.

Fibers

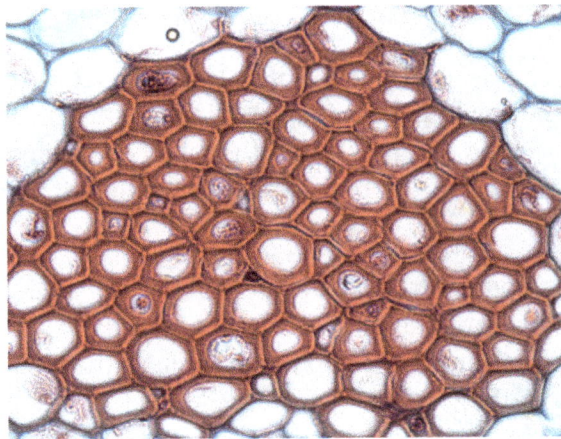

Cross section of sclerenchyma fibres

Fibers or bast are generally long, slender, so-called prosenchymatous cells, usually occurring in strands or bundles. Such bundles or the totality of a stem's bundles are colloquially called fibres. Their high load-bearing capacity and the ease with which they can be processed has since antiquity made them the source material for a number of things, like ropes, fabrics and mattresses. The fibres of flax (*Linum usitatissimum*) have been known in Europe and Egypt for more than 3,000 years, those of hemp (*Cannabis sativa*) in China for just as long. These fibres, and those of jute (*Corchorus capsularis*) and ramie (*Boehmeria nivea*, a nettle), are extremely soft and elastic and are especially well suited for the processing to textiles. Their principal cell wall material is cellulose.

Contrasting are hard fibres that are mostly found in monocots. Typical examples are the fibres of many grasses, agaves (sisal: *Agave sisalana*), lilies (*Yucca* or *Phormium tenax*), *Musa textilis* and others. Their cell walls contain, besides cellulose, a high proportion of lignin. The load-bearing capacity of *Phormium tenax* is as high as 20–25 kg/ mm^2, the same as that of good steel wire (25 kg/ mm^2), but the fibre tears as soon as too great a strain is placed upon it, while the wire distorts and does not tear before a strain of 80 kg/mm^2. The thickening of a cell wall has been studied in *Linum*. Starting at the centre of the fibre, the thickening layers of the secondary wall are deposited one after the other. Growth at both tips of the cell leads to simultaneous elongation. During development the layers of secondary material seem like tubes, of which the outer one is always longer and older than the next. After completion of growth, the missing parts are supplemented, so that the wall is evenly thickened up to the tips of the fibres.

Fibres usually originate from meristematic tissues. Cambium and procambium are their main centres of production. They are usually associated with the xylem and phloem of the vascular bundles. The fibres of the xylem are always lignified, while those of the phloem are cellulosic. Reliable evidence for the fibre cells' evolutionary origin from tracheids exists. During evolution the strength of the tracheid cell walls was enhanced, the ability to conduct water was lost and the size of the pits was reduced. Fibres that do not belong to the xylem are bast (outside the ring of cambium) and such fibres that are arranged in characteristic patterns at different sites of the shoot.

Sclereids

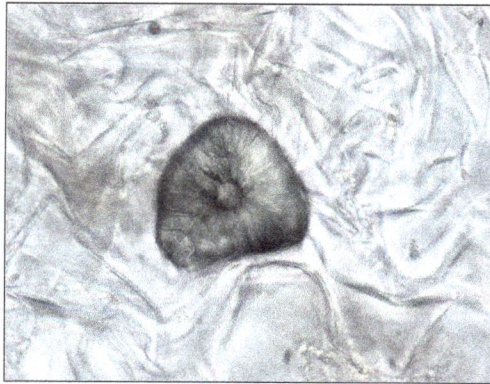

Fresh mount of a sclereid

Sclereids are a reduced form of sclerenchyma cells with highly thickened, lignified walls. These have a shape of a star.

They are small bundles of sclerenchyma tissue in plants that form durable layers, such as the cores of apples and the gritty texture of pears (*Pyrus communis*). Sclereids are variable in shape. The cells can be isodiametric, prosenchymatic, forked or elaborately branched. They can be grouped into bundles, can form complete tubes located at the periphery or can occur as single cells or small groups of cells within parenchyma tissues. But compared with most fibres, sclereids are relatively short. Characteristic examples are brachysclereids or the stone cells (called stone cells because of their hardness) of pears and quinces (*Cydonia oblonga*) and those of the shoot of the wax plant

(*Hoya carnosa*). The cell walls fill nearly all the cell's volume. A layering of the walls and the existence of branched pits is clearly visible. Branched pits such as these are called ramiform pits. The shell of many seeds like those of nuts as well as the stones of drupes like cherries and plums are made up from sclereids.

Long, tapered sclereids supporting a leaf edge in *Dionysia kossinskyi*

These structures are used to protect other cells.

Plant Cell Culture

Cell culture in a special tissue culture dish

Cell culture is the process by which cells are grown under controlled conditions, generally outside of their natural environment. Cell culture conditions can vary for each cell type, but artificial environments consist of a suitable vessel with substrate or medium that supplies the essential nutrients (amino acids, carbohydrates, vitamins, minerals), growth factors, hormones, and gases (CO_2, O_2), and regulates the physio-chemical environment (pH buffer, osmotic pressure, temperature). Most cells require a surface or an artificial substrate (adherent or monolayer culture) whereas others can be grown free floating in culture medium (suspension culture).

In practice, the term "cell culture" now refers to the culturing of cells derived from multicellular eukaryotes, especially animal cells, in contrast with other types of culture that also grow cells, such as plant tissue culture, fungal culture, and microbiological culture (of microbes). The historical development and methods of cell culture are closely interrelated to those of tissue culture and organ culture. Viral culture is also related, with cells as hosts for the viruses.

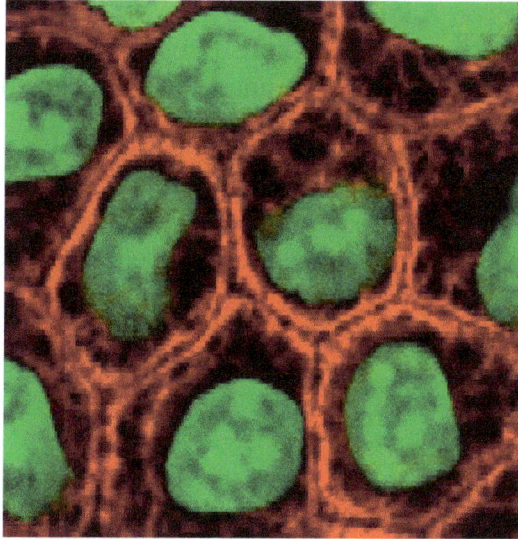

Epithelial cells in culture, stained for keratin (red) and DNA (green)

The laboratory technique of maintaining live cell lines (a population of cells descended from a single cell and containing the same genetic makeup) separated from their original tissue source became more robust in the middle 20th century.

History

The 19th-century English physiologist Sydney Ringer developed salt solutions containing the chlorides of sodium, potassium, calcium and magnesium suitable for maintaining the beating of an isolated animal heart outside of the body. In 1885, Wilhelm Roux removed a portion of the medullary plate of an embryonic chicken and maintained it in a warm saline solution for several days, establishing the principle of tissue culture. Ross Granville Harrison, working at Johns Hopkins Medical School and then at Yale University, published results of his experiments from 1907 to 1910, establishing the methodology of tissue culture.

Cell culture techniques were advanced significantly in the 1940s and 1950s to support research in virology. Growing viruses in cell cultures allowed preparation of purified viruses for the manufacture of vaccines. The injectable polio vaccine developed by Jonas Salk was one of the first products mass-produced using cell culture techniques. This vaccine was made possible by the cell culture research of John Franklin Enders, Thomas Huckle Weller, and Frederick Chapman Robbins, who were awarded a Nobel Prize for their discovery of a method of growing the virus in monkey kidney cell cultures.

Concepts in Mammalian Cell Culture

Isolation of Cells

Cells can be isolated from tissues for *ex vivo* culture in several ways. Cells can be easily purified from blood; however, only the white cells are capable of growth in culture. Mononuclear cells can be released from soft tissues by enzymatic digestion with enzymes such as collagenase, trypsin, or pronase, which break down the extracellular matrix. Alternatively, pieces of tissue can be placed in growth media, and the cells that grow out are available for culture. This method is known as explant culture.

Cells that are cultured directly from a subject are known as primary cells. With the exception of some derived from tumors, most primary cell cultures have limited lifespan.

An established or immortalized cell line has acquired the ability to proliferate indefinitely either through random mutation or deliberate modification, such as artificial expression of the telomerase gene. Numerous cell lines are well established as representative of particular cell types.

Maintaining Cells in Culture

For the majority of isolated primary cells, they undergo the process of senescence and stop dividing after a certain number of population doublings while generally retaining their viability (described as the Hayflick limit).

Cells are grown and maintained at an appropriate temperature and gas mixture (typically, 37 °C, 5% CO_2 for mammalian cells) in a cell incubator. Culture conditions vary widely for each cell type, and variation of conditions for a particular cell type can result in different phenotypes.

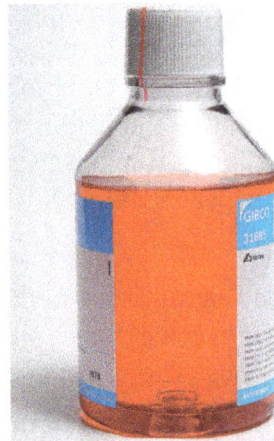

A bottle of DMEM cell culture medium

Aside from temperature and gas mixture, the most commonly varied factor in culture systems is the cell growth medium. Recipes for growth media can vary in pH, glucose concentration, growth factors, and the presence of other nutrients. The growth factors used to supplement media are often derived from the serum of animal blood, such as fetal bovine serum (FBS), bovine calf serum, equine serum, and porcine serum. One complication of these blood-derived ingredients is the potential for contamination of the culture with viruses or prions, particularly in medical biotechnology applications. Current practice is to minimize or eliminate the use of these ingredients wherever possible and use human platelet lysate (hPL). This eliminates the worry of cross-species contamination when using FBS with human cells. hPL has emerged as a safe and reliable alternative as a direct replacement for FBS or other animal serum. In addition, chemically defined media can be used to eliminate any serum trace (human or animal), but this cannot always be accomplished with different cell types. Alternative strategies involve sourcing the animal blood from countries with minimum BSE/TSE risk, such as The United States, Australia and New Zealand, and using purified nutrient concentrates derived from serum in place of whole animal serum for cell culture.

Plating density (number of cells per volume of culture medium) plays a critical role for some cell types. For example, a lower plating density makes granulosa cells exhibit estrogen production, while a higher plating density makes them appear as progesterone-producing theca lutein cells.

Cells can be grown either in suspension or adherent cultures. Some cells naturally live in suspension, without being attached to a surface, such as cells that exist in the bloodstream. There are also cell lines that have been modified to be able to survive in suspension cultures so they can be grown to a higher density than adherent conditions would allow. Adherent cells require a surface, such as tissue culture plastic or microcarrier, which may be coated with extracellular matrix (such as collagen and laminin) components to increase adhesion properties and provide other signals needed for growth and differentiation. Most cells derived from solid tissues are adherent. Another type of adherent culture is organotypic culture, which involves growing cells in a three-dimensional (3-D) environment as opposed to two-dimensional culture dishes. This 3D culture system is biochemically and physiologically more similar to *in vivo* tissue, but is technically challenging to maintain because of many factors (e.g. diffusion).

Components of Cell Culture Media

Component	Function
Carbon source (glucose/ glutamine)	Source of energy
Amino acid	Building blocks of protein
Vitamins	Promote cell survival and growth
Balanced salt solution	An isotonic mixture of ions to maintain optimum osmotic pressure within the cells and provide essential metal ions to act as cofactors for enzymatic reactions, cell adhesion etc.

| Phenol red dye | pH indicator. The color of phenol red changes from orange/red at pH 7-7.4 to yellow at acidic (lower) pH and purple at basic (higher) pH. |
| Bicarbonate /HEPES buffer | It is used to maintain a balanced pH in the media |

Growth Conditions

Parameter	
Temperature	37 °C
CO2	5%
Humidity	95%

Cell Line Cross-contamination

Cell line cross-contamination can be a problem for scientists working with cultured cells. Studies suggest anywhere from 15–20% of the time, cells used in experiments have been misidentified or contaminated with another cell line. Problems with cell line cross-contamination have even been detected in lines from the NCI-60 panel, which are used routinely for drug-screening studies. Major cell line repositories, including the American Type Culture Collection (ATCC), the European Collection of Cell Cultures (ECACC) and the German Collection of Microorganisms and Cell Cultures (DSMZ), have received cell line submissions from researchers that were misidentified by them. Such contamination poses a problem for the quality of research produced using cell culture lines, and the major repositories are now authenticating all cell line submissions. ATCC uses short tandem repeat (STR) DNA fingerprinting to authenticate its cell lines.

To address this problem of cell line cross-contamination, researchers are encouraged to authenticate their cell lines at an early passage to establish the identity of the cell line. Authentication should be repeated before freezing cell line stocks, every two months during active culturing and before any publication of research data generated using the cell lines. Many methods are used to identify cell lines, including isoenzyme analysis, human lymphocyte antigen (HLA) typing, chromosomal analysis, karyotyping, morphology and STR analysis.

One significant cell-line cross contaminant is the immortal HeLa cell line.

Other Technical Issues

As cells generally continue to divide in culture, they generally grow to fill the available area or volume. This can generate several issues:

- Nutrient depletion in the growth media.

- Changes in pH of the growth media.

- Accumulation of apoptotic/necrotic (dead) cells.

- Cell-to-cell contact can stimulate cell cycle arrest, causing cells to stop dividing, known as contact inhibition.

- Cell-to-cell contact can stimulate cellular differentiation.

- Genetic and epigenetic alterations, with a natural selection of the altered cells potentially leading to overgrowth of abnormal, culture-adapted cells with decreased differentiation and increased proliferative capacity.

Manipulation of Cultured Cells

Among the common manipulations carried out on culture cells are media changes, passaging cells, and transfecting cells. These are generally performed using tissue culture methods that rely on aseptic technique. Aseptic technique aims to avoid contamination with bacteria, yeast, or other cell lines. Manipulations are typically carried out in a biosafety hood or laminar flow cabinet to exclude contaminating micro-organisms. Antibiotics (e.g. penicillin and streptomycin) and antifungals (e.g.amphotericin B) can also be added to the growth media.

As cells undergo metabolic processes, acid is produced and the pH decreases. Often, a pH indicator is added to the medium to measure nutrient depletion.

Media Changes

In the case of adherent cultures, the media can be removed directly by aspiration, and then is replaced. Media changes in non-adherent cultures involve centrifuging the culture and resuspending the cells in fresh media.

Passaging Cells

Passaging (also known as subculture or splitting cells) involves transferring a small number of cells into a new vessel. Cells can be cultured for a longer time if they are split regularly, as it avoids the senescence associated with prolonged high cell density. Suspension cultures are easily passaged with a small amount of culture containing a few cells diluted in a larger volume of fresh media. For adherent cultures, cells first need to be detached; this is commonly done with a mixture of trypsin-EDTA; however, other enzyme mixes are now available for this purpose. A small number of detached cells can then be used to seed a new culture. Some cell cultures, such as RAW cells are mechanically scraped from the surface of their vessel with rubber scrapers.

Transfection and Transduction

Another common method for manipulating cells involves the introduction of foreign DNA by transfection. This is often performed to cause cells to express a gene of interest. More recently, the transfection of RNAi constructs have been realized as a convenient mechanism for suppressing the expression of a particular gene/protein. DNA can also be inserted into cells using viruses, in methods referred to as transduction, infection or transformation. Viruses, as parasitic agents, are well suited to introducing DNA into cells, as this is a part of their normal course of reproduction.

Established Human Cell Lines

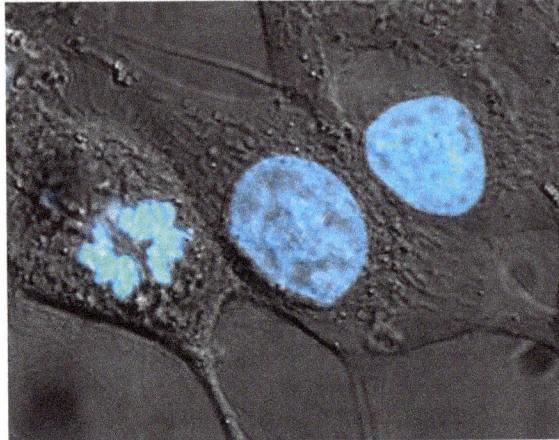

Cultured HeLa cells have been stained with Hoechst turning their nuclei blue, and are one of the earliest human cell lines descended from Henrietta Lacks, who died of cervical cancer from which these cells originated.

Cell lines that originate with humans have been somewhat controversial in bioethics, as they may outlive their parent organism and later be used in the discovery of lucrative medical treatments. In the pioneering decision in this area, the Supreme Court of California held in *Moore v. Regents of the University of California* that human patients have no property rights in cell lines derived from organs removed with their consent.

It is possible to fuse normal cells with an immortalised cell line. This method is used to produce monoclonal antibodies. In brief, lymphocytes isolated from the spleen (or possibly blood) of an immunised animal are combined with an immortal myeloma cell line (B cell lineage) to produce a hybridoma which has the antibody specificity of the primary lymphocyte and the immortality of the myeloma. Selective growth medium (HA or HAT) is used to select against unfused myeloma cells; primary lymphoctyes die quickly in culture and only the fused cells survive. These are screened for production of the required antibody, generally in pools to start with and then after single cloning.

Cell Strains

A cell strain is derived either from a primary culture or a cell line by the selection or cloning of cells having specific properties or characteristics which must be defined. Cell strains are cells that have been adapted to culture but, unlike cell lines, have a finite division potential. Non-immortalized cells stop dividing after 40 to 60 population doublings and, after this, they lose their ability to proliferate (a genetically determined event known as senescence).

Applications of Cell Culture

Mass culture of animal cell lines is fundamental to the manufacture of viral vaccines and other products of biotechnology. Culture of human stem cells is used to expand the number of cells and differentiate the cells into various somatic cell types for transplantation. Stem cell culture is also used to harvest the molecules and exosomes that the stem cells release for the purposes of therapeutic development.

Biological products produced by recombinant DNA (rDNA) technology in animal cell cultures include enzymes, synthetic hormones, immunobiologicals (monoclonal antibodies, interleukins, lymphokines), and anticancer agents. Although many simpler proteins can be produced using rDNA in bacterial cultures, more complex proteins that are glycosylated (carbohydrate-modified) currently must be made in animal cells. An important example of such a complex protein is the hormone erythropoietin. The cost of growing mammalian cell cultures is high, so research is underway to produce such complex proteins in insect cells or in higher plants, use of single embryonic cell and somatic embryos as a source for direct gene transfer via particle bombardment, transit gene expression and confocal microscopy observation is one of its applications. It also offers to confirm single cell origin of somatic embryos and the asymmetry of the first cell division, which starts the process.

Cell culture is also a key technique for cellular agriculture, which aims to provide both new products and new ways of producing existing agricultural products like milk, (cultured) meat, fragrances, and rhino horn from cells and microorganisms. It is therefore considered one means of achieving animal-free agriculture.

Cell Culture in Two Dimensions

Research in tissue engineering, stem cells and molecular biology primarily involves cultures of cells on flat plastic dishes. This technique is known as two-dimensional (2D) cell culture, and was first developed by Wilhelm Roux who, in 1885, removed a portion of the medullary plate of an embryonic chicken and maintained it in warm saline for several days on a flat glass plate. From the advance of polymer technology arose today's standard plastic dish for 2D cell culture, commonly known as the Petri dish. Julius Richard Petri, a German bacteriologist, is generally credited with this invention while working as an assistant to Robert Koch. Various researchers today also utilize culturing laboratory flasks, conicals, and even disposable bags like those used in single-use bioreactors.

Aside from Petri dishes, scientists have long been growing cells within biologically derived matrices such as collagen or fibrin, and more recently, on synthetic hydrogels such as polyacrylamide or PEG. They do this in order to elicit phenotypes that are not expressed on conventionally rigid substrates. There is growing interest in controlling matrix stiffness, a concept that has led to discoveries in fields such as:

- Stem cell self-renewal
- Lineage specification
- Cancer cell phenotype
- Fibrosis
- Hepatocyte function
- Mechanosensing

Cell Culture in Three Dimensions

Cell culture in three dimensions has been touted as "Biology's New Dimension". At present, the practice of cell culture remains based on varying combinations of single or multiple cell structures in 2D. Currently, there is an increase in use of 3D cell cultures in research areas including drug discovery, cancer biology, regenerative medicine and basic life science research. 3D cell cultures can be grown using a scaffold or matrix, or in a scaffold-free manner. Scaffold based cultures utilize an acellular 3D matrix or a liquid matrix. Scaffold-free methods are normally generated in suspensions. There are a variety of platforms used to facilitate the growth of three-dimensional cellular structures including scaffold systems such as hydrogel matrices and solid scaffolds, and scaffold-free systems such as low-adhesion plates, nanoparticle facilitated magnetic levitation, and hanging drop plates.

3D Cell Culture in Scaffolds

Eric Simon, in a 1988 NIH SBIR grant report, showed that electrospinning could be used to produced nano- and submicron-scale polystyrene and polycarbonate fibrous scaffolds specifically intended for use as *in vitro* cell substrates. This early use of electrospun fibrous lattices for cell culture and tissue engineering showed that various cell types including Human Foreskin Fibroblasts (HFF), transformed Human Carcinoma (HEp-2), and Mink Lung Epithelium (MLE) would adhere to and proliferate upon polycarbonate fibers. It was noted that, as opposed to the flattened morphology typically seen in 2D culture, cells grown on the electrospun fibers exhibited a more histotypic rounded 3-dimensional morphology generally observed *in vivo*.

3D Cell Culture in Hydrogels

As the natural extracellular matrix (ECM) is important in the survival, proliferation, differentiation and migration of cells, different hydrogel culture matrices mimicking natural ECM structure are seen as potential approaches to in vivo –like cell culturing.

Hydrogels are composed of interconnected pores with high water retention, which enables efficient transport of substances such as nutrients and gases. Several different types of hydrogels from natural and synthetic materials are available for 3D cell culture, including animal ECM extract hydrogels, protein hydrogels, peptide hydrogels, polymer hydrogels, and wood-based nanocellulose hydrogel.

3D Cell Culturing by Magnetic Levitation

The 3D Cell Culturing by Magnetic Levitation method (MLM) is the application of growing 3D tissue by inducing cells treated with magnetic nanoparticle assemblies in spatially varying magnetic fields using neodymium magnetic drivers and promoting cell to cell interactions by levitating the cells up to the air/liquid interface of a standard petri dish. The magnetic nanoparticle assemblies consist of magnetic iron oxide nanoparticles, gold nanoparticles, and the polymer polylysine. 3D cell culturing is scalable, with the capability for culturing 500 cells to millions of cells or from single dish to high-throughput low volume systems.

Tissue Culture and Engineering

Cell culture is a fundamental component of tissue culture and tissue engineering, as it establishes the basics of growing and maintaining cells *in vitro*. The major application of human cell culture is in stem cell industry, where mesenchymal stem cells can be cultured and cryopreserved for future use. Tissue engineering potentially offers dramatic improvements in low cost medical care for hundreds of thousands of patients annually.

Vaccines

Vaccines for polio, measles, mumps, rubella, and chickenpox are currently made in cell cultures. Due to the H5N1 pandemic threat, research into using cell culture for influenza vaccines is being funded by the United States government. Novel ideas in the field include recombinant DNA-based vaccines, such as one made using human adenovirus (a common cold virus) as a vector, and novel adjuvants.

Culture of Non-mammalian Cells

Plant Cell Culture Methods

Plant cell cultures are typically grown as cell suspension cultures in a liquid medium or as callus cultures on a solid medium. The culturing of undifferentiated plant cells and calli requires the proper balance of the plant growth hormones auxin and cytokinin.

Insect Cell Culture

Cells derived from Drosophila melanogaster (most prominently, Schneider 2 cells) can be used for experiments which may be hard to do on live flies or larvae, such as biochemical studies or studies using siRNA. Cell lines derived from the army worm *Spodoptera frugiperda*, including Sf9 and Sf21, and from the cabbage looper *Trichoplusia ni*, High Five cells, are commonly used for expression of recombinant proteins using baculovirus.

Bacterial and Yeast Culture Methods

For bacteria and yeasts, small quantities of cells are usually grown on a solid support that contains nutrients embedded in it, usually a gel such as agar, while large-scale cultures are grown with the cells suspended in a nutrient broth.

Viral Culture Methods

The culture of viruses requires the culture of cells of mammalian, plant, fungal or bacterial origin as hosts for the growth and replication of the virus. Whole wild type viruses, recombinant viruses or viral products may be generated in cell types other than their natural hosts under the right conditions. Depending on the species of the virus, infection and viral replication may result in host cell lysis and formation of a viral plaque.

Common Cell Lines

Human cell lines

- DU145 (prostate cancer)

- H295R (adrenocortical cancer)

- HeLa (cervical cancer)

- KBM-7 (chronic myelogenous leukemia)

- LNCaP (prostate cancer)

- MCF-7 (breast cancer)

- MDA-MB-468 (breast cancer)

- PC3 (prostate cancer)

- SaOS-2 (bone cancer)

- SH-SY5Y (neuroblastoma, cloned from a myeloma)

- T47D (breast cancer)

- THP-1 (acute myeloid leukemia)

- U87 (glioblastoma)

- National Cancer Institute's 60 cancer cell line panel (NCI60)

Primate cell lines

- Vero (African green monkey *Chlorocebus* kidney epithelial cell line)

Mouse cell lines

- MC3T3 (embryonic calvarium)

Rat tumor cell lines

- GH3 (pituitary tumor)
- PC12 (pheochromocytoma)

Plant cell lines

- Tobacco BY-2 cells (kept as cell suspension culture, they are model system of plant cell)

Other species cell lines

- Dog MDCK kidney epithelial
- Xenopus A6 kidney epithelial
- Zebrafish AB9

Vascular Cambium

The vascular cambium (also called main cambium, wood cambium, bifacial cambium; plural *cambia*) is a plant tissue located between the xylem and the phloem in the stems and roots of vascular plants. It is a cylinder of unspecialized meristem cells that divide to form secondary vascular tissues. It is the source of both secondary xylem growth inwards towards the pith, and secondary phloem growth outwards to the bark. Unlike the xylem and phloem, it does not transport water, minerals or food through the plant.

Vascular cambia are found in dicots and gymnosperms but not monocots, which usually lack secondary growth. A few leaf types also have a vascular cambium. In wood, the vascular cambium is the obvious line separating the bark and wood. For successful grafting, the vascular cambia of the rootstock and scion must be aligned so they can grow together.

Structure and Function

The cambium present between primary xylem and primary phloem is called *intrafasicular* cambium. During secondary growth, cells of meduallary rays, in a line with intrafasicular cambium, become meristematic and form *interfascicular* cambium. Therefore, the intrafascicular and interfascicular cambia form a ring which separates the primary xylem and primary phloem, and is known as *cambium ring*. The vascular cambium produces secondary xylem on the inside of the ring, and secondary phloem on the outside, pushing the primary xylem and phloem apart.

The vascular cambium usually consists of two types of cells:

- Fusiform initials (tall, axially oriented)
- Ray initials (smaller and round to angular in shape)

Maintenance of Cambial Meristem

The vascular cambium is maintained by a network of interacting signal feedback loops. Currently, both hormones and short peptides have been identified as information carriers in these systems. While similar regulation occurs in other meristems of plants, the cambial meristem receives signals from both the xylem and phloem sides for the meristem. Signals received from outside the meristem act to down regulate internal factors, which promotes cell proliferation, and promotes differentiation.

Hormonal Regulation

The phytohormones that are involved in the vascular cambial activity are auxins, ethylene, gibberellins, cytokinins, abscisic acid and more to be discovered. Each one of these plant hormones are vital for the regulation of the cambial activity and are dependent on their concentration.

Auxin hormones are proven to stimulate mitoses, cell production and regulate interfascicular and fascicular cambium. Applying auxin to the surface of a tree stump allowed decapitated shoots to continue secondary growth. The absence of auxin hormones will have a detrimental effect on a plant. It has been shown that mutants without auxin will exhibit increased spacing between the interfascicular cambiums and reduced growth of the vascular bundles. The mutant plant will therefore experience a decreased in water, nutrients, and photosynthates being transported throughout the plant, eventually leading to death. Auxin also regulates the two types of cell in the vascular cambium, ray and fusiform initials. Regulation of these initials ensures the connection and communication between xylem and phloem is maintained for the translocation of nourishment and sugars are safely being stored as an energy resource. Ethylene levels are high in plants with an active cambial zone and are still currently being studied. Gibberellin stimulates the cambial cell division and also regulates differentiation of the xylem tissues, with no effect on the rate of phloem differentiation. Differentiation is an essential process that changes these tissues into a more specialized type, leading to an important role in maintaining the life form of a plant. In poplar trees, high concentrations of gibberellin is positively correlated to an increase of cambial cell division and an increase of auxin in the cambial stem cells. Gibberellin is also responsible for the expansion of xylem through a signal traveling from the shoot to the root. Cytokinin hormone is known to regulate the rate of the cell division instead of the direction of cell differentiation. A study demonstrated that the mutants are found to have a reduction in stem and root growth but the secondary vascular pattern of the vascular bundles were not affected with a treatment of cytokinin.

Transfer Cell

Transfer cells are specialized parenchyma cells that have an increased surface area, due to infoldings of the plasma membrane. They facilitate the transport of sugars from a sugar source, mainly mature leaves, to a sugar sink, often developing leaves or fruits. They are found in nectaries of flowers and some carnivorous plants. Transfer cells are specially found in plants in the region of absorption or secretion of nutrients.

The term transfer cell was coined by B. E. S. Gunning and J. S. Pate.

Sclereid

Sclereids are a reduced form of sclerenchyma cells with highly thickened, lignified cellular walls that form small bundles of durable layers of tissue in most plants. The presence of numerous sclereids form the cores of apples and produce the gritty texture of pears.

Although sclereids are variable in shape, the cells are generally isodiametric, prosenchymatic, forked or elaborately branched. They can be grouped into bundles, can form complete tubes located at the periphery or can occur as single cells or small groups of cells within parenchyma tissues.

When compared with most fibres, sclereids are relatively short. Characteristic examples are brachysclereids or the stone cells (called stone cells because of their hardness) of pears (*Pyrus communis*) and quinces (*Cydonia oblonga*) and those of the shoot of the wax plant (*Hoya carnosa*). The cell walls fill nearly all the cell's volume. A layering of the walls and the existence of branched pits is clearly visible. Branched pits such as these are called ramiform pits. The shell of many seeds like those of nuts as well as the stones of drupes like cherries or plums are made up from sclereids.

These structures are used to protect other cells.

Aerenchyma

Aerenchyma is a spongy tissue that forms spaces or air channels in the leaves, stems and roots of some plants, which allows exchange of gases between the shoot and the root. The channels of air-filled cavities provide a low-resistance internal pathway for the exchange of gases such as oxygen and ethylene between the plant above the water and the submerged tissues. Aerenchyma is also widespread in aquatic and wetland plants which must grow in hypoxic soils.

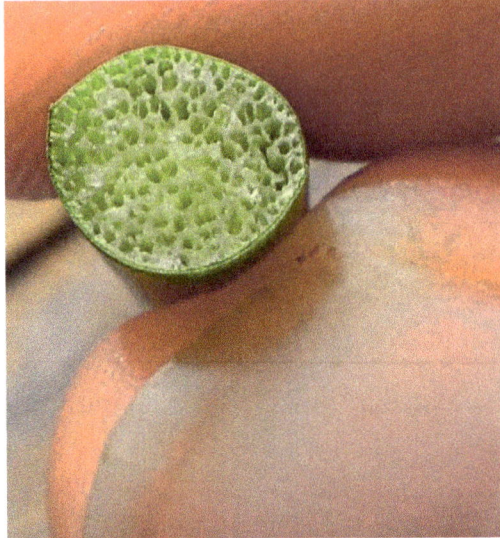

Aerenchyma in stem cross section of a typical wetland plant

Aerenchyma Formation and Hypoxia

When soil is flooded, hypoxia develops, as soil microorganisms consume oxygen faster than diffusion occurs. The presence of hypoxic soils is one of the defining characteristics of wetlands. Many wetland plants possess aerenchyma, and in some, such as water-lilies, there is mass flow of atmospheric air through leaves and rhizomes. There are many other chemical consequences of hypoxia. For example, nitrification is inhibited as low oxygen occurs and toxic compounds are formed, as anaerobic bacteria use nitrate, manganese, and sulfate as alternative electron acceptors. The reduction-oxidation potential of the rhizhosphere decreases and metal ions such as iron and manganese precipitate. Aerenchyma is a modification of the parenchyma.

In general, low oxygen stimulates trees and plants to produce ethylene. Ethylene slows down primary and adventitious root elongation and formation.

Advantages of Aerenchyma

The large air-filled cavities provide a low-resistance internal pathway for the exchange of gases between the plant organs above the water and the submerged tissues. This allows plants to grow without incurring the metabolic costs of anaerobic respiration. Some of the oxygen transported through the aerenchyma leaks through root pores into the surrounding soil. The resulting small rhizosphere of oxygenated soil around individual roots support microorganisms that prevent the influx of potentially toxic soil components such as sulfide, iron, and manganese.

Vascular Tissue

Cross section of celery stalk, showing vascular bundles, which include both phloem and xylem

Vascular tissue is a complex conducting tissue, formed of more than one cell type, found in vascular plants. The primary components of vascular tissue are the xylem and phloem. These two tissues transport fluid and nutrients internally. There are also two meristems associated with vascular tissue: the vascular cambium and the cork cambium. All the vascular tissues within a particular plant together constitute the vascular tissue system of that plant.

Detail of the vasculature of a brambleleaf

The cells in vascular tissue are typically long and slender. Since the xylem and phloem function in the conduction of water, minerals, and nutrients throughout the plant, it is not surprising that their form should be similar to pipes. The individual cells of phloem are connected end-to-end, just as the sections of a pipe might be. As the plant grows, new vascular tissue differentiates in the growing tips of the plant. The new tissue is aligned with existing vascular tissue, maintaining its connection throughout the plant. The vascular tissue in plants is arranged in long, discrete strands called vascular bundles. These bundles include both xylem and phloem, as well as supporting and protective cells. In stems and roots, the xylem typically lies closer to the interior of the stem with phloem towards the exterior of the stem. In the stems of some Asterales dicots, there may be phloem located inwardly from the xylem as well.

Between the xylem and phloem is a meristem called the vascular cambium. This tissue divides off cells that will become additional xylem and phloem. This growth increases the girth of the plant, rather than its length. As long as the vascular cambium continues to produce new cells, the plant will continue to grow more stout. In trees and other plants that develop wood, the vascular cambium allows the expansion of vascular tissue that produces woody growth. Because this growth ruptures the epidermis of the stem, woody plants also have a cork cambium that develops among the phloem. The cork cambium gives rise to thickened cork cells to protect the surface of the plant and reduce water loss. Both the production of wood and the production of cork are forms of secondary growth.

In leaves, the vascular bundles are located among the spongy mesophyll. The xylem is oriented toward the adaxial surface of the leaf (usually the upper side), and phloem is oriented toward the abaxial surface of the leaf. This is why aphids are typically found on the underside of the leaves rather than on the top, since the phloem transports sugars manufactured by the plant and they are closer to the lower surface.

Vascular Bundle

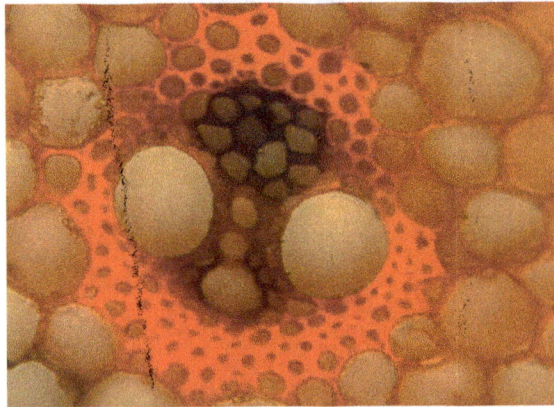

Detail of vascular bundle: closed, collateral vascular bundles of the stem axis of *Zea mays*

A vascular bundle is a part of the transport system in vascular plants. The transport itself happens in vascular tissue, which exists in two forms: xylem and phloem. Both these tissues are present in a vascular bundle, which in addition will include supporting and protective tissues.

The xylem typically lies adaxial with phloem positioned abaxial. In a stem or root this means that the xylem is closer to the centre of the stem or root while the phloem is closer to the exterior. In a leaf, the adaxial surface of the leaf will usually be the upper side, with the abaxial surface the lower side. This is why aphids are typically found on the underside of a leaf rather than on the top, since the sugars manufactured by the plant are transported by the phloem, which is closer to the lower surface.

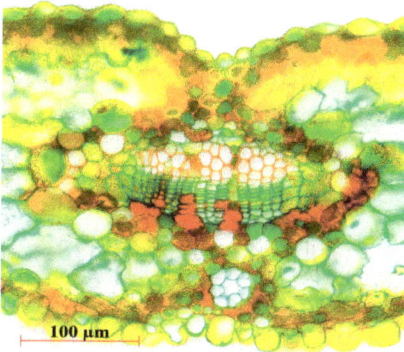

Vascular bundle in the leaf of *Metasequoia glyptostroboides*.

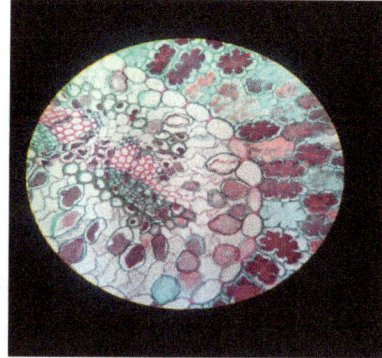

The vascular bundle of pine leaf showing xylem and phloem.

The position of vascular bundles relative to each other may vary considerably.

Bundle-sheath Cells

Bundle-sheath cells are photosynthetic cells arranged into tightly packed sheaths around the veins of a leaf. They form a protective covering on leaf veins, and consist of one or more cell layers, usually parenchyma. Loosely arranged mesophyll cells lie between the bundle sheath and the leaf surface. The Calvin cycle is confined to the chloroplasts of these bundle sheath cells in C_4 plants.

Meristem

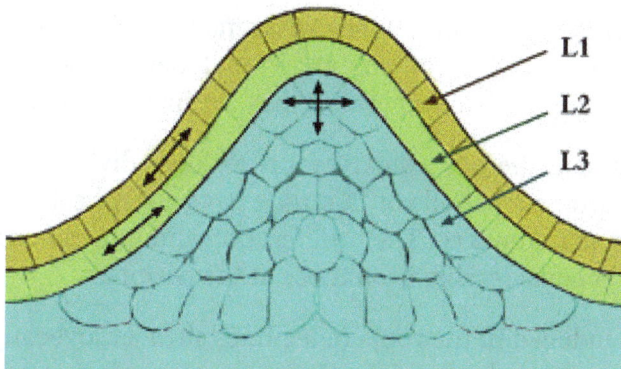

Tunica-Corpus model of the apical meristem (growing tip). The epidermal (L1) and subepidermal (L2) layers form the outer layers called the tunica. The inner L3 layer is called the corpus. Cells in the L1 and L2 layers divide in a sideways fashion, which keeps these layers distinct, whereas the L3 layer divides in a more random fashion.

A meristem is the tissue in most plants containing undifferentiated cells (meristematic cells), found in zones of the plant where growth can take place.

Meristematic cells give rise to various organs of the plant and keep the plant growing. The *shoot apical meristem* (SAM) gives rise to organs like the leaves and flowers, while the *root apical meristem* (RAM) provides the meristematic cells for the future root growth. SAM and RAM cells divide rapidly and are considered indeterminate, in that they do not possess any defined end status. In that sense, the meristematic cells are frequently compared to the stem cells in animals, which have an analogous behavior and function.

The term *meristem* was first used in 1858 by Karl Wilhelm von Nägeli (1817–1891) in his book *Beiträge zur Wissenschaftlichen Botanik* ("Contributions to Scientific Botany"). It is derived from the Greek word *merizein*, meaning to divide, in recognition of its inherent function.

In general, differentiated plant cells cannot divide or produce cells of a different type. Therefore, cell division in the meristem is required to provide new cells for expansion and differentiation of tissues and initiation of new organs, providing the basic structure of the plant body.

Meristematic cells are incompletely or not at all differentiated, and are capable of continued cellular division (youthful). Furthermore, the cells are small and protoplasm fills the cell completely. The vacuoles are extremely small. The cytoplasm does not contain differentiated plastids (chloroplasts or chromoplasts), although they are present in rudimentary form (proplastids). Meristematic cells are packed closely together without intercellular cavities. The cell wall is a very thin *primary cell wall*.

Maintenance of the cells requires a balance between two antagonistic processes: organ initiation and stem cell population renewal.

Apical meristems are the completely undifferentiated (indeterminate) meristems in a plant. These differentiate into three kinds of primary meristems. The primary meristems in turn produce the two secondary meristem types. These secondary meristems are also known as lateral meristems because they are involved in lateral growth.

At the meristem summit, there is a small group of slowly dividing cells, which is commonly called the central zone. Cells of this zone have a stem cell function and are essential for meristem maintenance. The proliferation and growth rates at the meristem summit usually differ considerably from those at the periphery.

Meristems also are induced in the roots of legumes such as soybean, *Lotus japonicus*, pea, and *Medicago truncatula* after infection with soil bacteria commonly called Rhizobium. Cells of the inner or outer cortex in the so-called "window of nodulation" just behind the developing root tip are induced to divide. The critical signal substance is the lipo-oligosaccharide Nod-factor, decorated with side groups to allow specificity of interaction. The Nod factor receptor proteins NFR1 and NFR5 were cloned from several legumes including *Lotus japonicus*, *Medicago truncatula* and soybean (*Glycine max*). Regulation of nodule meristems utilizes long distance regulation commonly called "Autoregulation of Nodulation" (AON). This process involves a leaf-vascular tissue located LRR receptor kinases (LjHAR1, GmNARK and MtSUNN), CLE peptide signalling, and KAPP interaction, similar to that seen in the CLV1,2,3 system. LjKLAVIER also exhibits a nodule regulation phenotype though it is not yet known how this relates to the other AON receptor kinases.

Apical Meristems

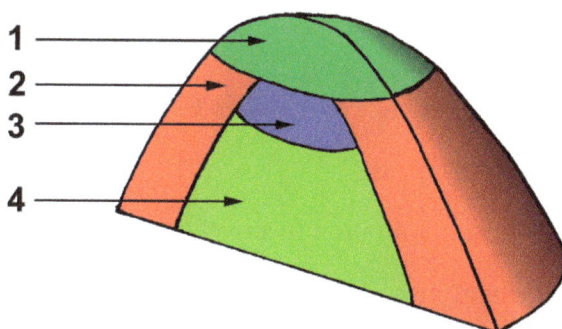

Organisation of an apical meristem (growing tip)
1 - Central zone 2 - Peripheral zone 3 - Medullary (i.e. central) meristem 4 - Medullary tissue

The number of layers varies according to plant type. In general the outermost layer is called the tunica while the innermost layers are the corpus. In monocots, the tunica determine the physical characteristics of the leaf edge and margin. In dicots, layer two of the corpus determine the characteristics of the edge of the leaf. The corpus and tunica play a critical part of the plant physical appearance as all plant cells are formed from the meristems. Apical meristems are found in two locations: the root and the stem. Some Arctic plants have an apical meristem in the lower/middle parts of the plant. It is thought that this kind of meristem evolved because it is advantageous in Arctic conditions.

Shoot Apical Meristems

Shoot apical meristems of *Crassula ovata* (left). Fourteen days later, leaves have developed (right).

The source of all above-ground organs. Cells at the shoot apical meristem summit serve as stem cells to the surrounding peripheral region, where they proliferate rapidly and are incorporated into differentiating leaf or flower primordia.

The shoot apical meristem is the site of most of the embryogenesis in flowering plants. Primordia of leaves, sepals, petals, stamens and ovaries are initiated here at the rate of one every time interval, called a plastochron. It is where the first indications that flower development has been evoked are manifested. One of these indications might be the loss of apical dominance and the release of otherwise dormant cells to develop as auxiliary shoot meristems, in some species in axils of primordia as close as two or three away from the apical dome. The shoot apical meristem consists of 4 distinct cell groups:

- Stem cells

- The immediate daughter cells of the stem cells

- A subjacent organising centre

- Founder cells for organ initiation in surrounding regions

The four distinct zones mentioned above are maintained by a complex signalling pathway. In *Arabidopsis thaliana*, 3 interacting *CLAVATA* genes are required to regulate the size of the stem cell reservoir in the shoot apical meristem by controlling the rate of cell division. CLV1 and CLV2 are predicted to form a receptor complex (of the LRR receptor-like kinase family) to which CLV3 is a ligand. CLV3 shares some homology with the ESR proteins of maize, with a short 14 amino acid region being conserved between the proteins. Proteins that contain these conserved regions have been grouped into the CLE family of proteins.

CLV1 has been shown to interact with several cytoplasmic proteins that are most likely involved in downstream signalling. For example, the CLV complex has been found to be associated with Rho/Rac small GTPase-related proteins. These proteins may act as an intermediate between the CLV complex and a mitogen-activated protein kinase (MAPK), which is often involved in signalling cascades. KAPP is a kinase-associated protein phosphatase that has been shown to interact with CLV1. KAPP is thought to act as a negative regulator of CLV1 by dephosphorylating it.

Another important gene in plant meristem maintenance is *WUSCHEL* (shortened to *WUS*), which is a target of CLV signalling. *WUS* is expressed in the cells below the stem cells of the meristem and its presence prevents the differentiation of the stem cells. CLV1 acts to promote cellular differentiation by repressing *WUS* activity outside of the central zone containing the stem cells. *SHOOT MERISTEMLESS* (*STM*) also acts to prevent the differentiation of stem cells by repressing the expression of MYB genes that are involved in cellular differentiation.

Root Apical Meristem

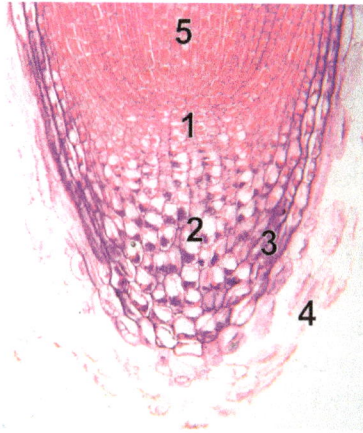

10x microscope image of root tip with meristem
1 - quiescent center 2 - calyptrogen (live rootcap cells) 3 - rootcap
4 - sloughed off dead rootcap cells 5 - procambium

Unlike the shoot apical meristem, the root apical meristem produces cells in two dimensions. It harbors two pools of stem cells around an organizing center called the quiescent center (QC) cells and together produce most of the cells in an adult root. At its apex, the root meristem is covered by the root cap, which protects and guides its growth trajectory. Cells are continuously sloughed off the outer surface of the root cap. The QC cells are characterized by their low mitotic activity. Evidence suggests that the QC maintains the surrounding stem cells by preventing their differentiation, via signal(s) that are yet to be discovered. This allows a constant supply of new cells in the meristem required for continuous root growth. Recent findings indicate that QC can also act as a reservoir of stem cells to replenish whatever is lost or damaged. Root apical meristem and tissue patterns become established in the embryo in the case of the primary root, and in the new lateral root primordium in the case of secondary roots.

Intercalary Meristem

In angiosperms, intercalary meristems occur only in monocot (in particular, grass) stems at the base of nodes and leaf blades. Horsetails also exhibit intercalary growth. Intercalary meristems are capable of cell division, and they allow for rapid growth and regrowth of many monocots. Intercalary meristems at the nodes of bamboo allow for rapid stem elongation, while those at the base of most grass leaf blades allow damaged leaves to rapidly regrow. This leaf regrowth in grasses evolved in response to damage by grazing herbivores, but is more familiar to us in response to lawnmowers.

Floral Meristem

When plants begin the developmental process known as flowering, the shoot apical meristem is transformed into an inflorescence meristem, which goes on to produce the floral meristem, which produces the sepals, petals, stamens, and carpels of the flower.

In contrast to vegetative apical meristems and some exflorescence meristems, floral meristems cannot continue to grow indefinitely. Their future growth is limited to the flower with a particular size and form. The transition from shoot meristem to floral meristem requires floral meristem identity genes, that both specify the floral organs and cause the termination of the production of stem cells. *AGAMOUS* (*AG*) is a floral homeotic gene required for floral meristem termination and necessary for proper development of the stamens and carpels. *AG* is necessary to prevent the conversion of floral meristems to inflorescence shoot meristems, but is not involved in the transition from shoot to floral meristem. AG is turned on by the floral meristem identity gene *LEAFY* (*LFY*) and *WUS* and is restricted to the centre of the floral meristem or the inner two whorls. This way floral identity and region specificity is achieved. WUS activates AG by binding to a consensus sequence in the AG's second intron and LFY binds to adjacent recognition sites. Once AG is activated it represses expression of WUS leading to the termination of the meristem.

Through the years, scientists have manipulated floral meristems for economic reasons. An example is the mutant tobacco plant "Maryland Mammoth." In 1936, the department of agriculture of Switzerland performed several scientific tests with this plant. "Maryland Mammoth" is peculiar in that it grows much faster than other tobacco plants.

Apical Dominance

Apical dominance is phenomenon where one meristem prevents or inhibits the growth of other meristems. As a result, the plant will have one clearly defined main trunk. For example, in trees, the tip of the main trunk bears the dominant meristem. Therefore, the tip of the trunk grows rapidly and is not shadowed by branches. If the dominant meristem is cut off, one or more branch tips will assume dominance. The branch will start growing faster and the new growth will be vertical. Over the years, the branch may begin to look more and more like an extension of the main trunk. Often several branches will exhibit this behaviour after the removal of apical meristem, leading to a bushy growth.

The mechanism of apical dominance is based on the plant hormone auxin. It is produced in the apical meristem and transported towards the roots in the cambium. If apical dominance is complete, it prevents any branches from forming as long as the apical meristem is active. If the dominance is incomplete, side branches will develop.

Recent investigations into apical dominance and the control of branching have revealed a new plant hormone family termed strigolactones. These compounds were previously known to be involved in seed germination and communication with mycorrhizal fungi and are now shown to be involved in inhibition of branching.

Diversity in Meristem Architectures

Is the mechanism of being *indeterminate* conserved in the SAMs of the plant world? The SAM contains a population of stem cells that also produce the lateral meristems while the stem elongates. It turns out that the mechanism of regulation of the stem cell number might indeed be evolutionarily conserved. The *CLAVATA* gene *CLV2* responsible for maintaining the stem cell population in *Arabidopsis thaliana* is very closely related to the Maize gene *FASCIATED EAR 2*(*FEA2*) also involved in the same function. Similarly, in Rice, the *FON1-FON2* system seems to bear a close relationship with the CLV signaling system in *Arabidopsis thaliana*. These studies suggest that the regulation of stem cell number, identity and differentiation might be an evolutionarily conserved mechanism in monocots, if not in angiosperms. Rice also contains another genetic system distinct from *FON1-FON2*, that is involved in regulating stem cell number. This example underlines the innovation that goes about in the living world all the time.

Role of the KNOX-family Genes

Genetic screens have identified genes belonging to the KNOX family in this function. These genes essentially maintain the stem cells in an undifferentiated state. The KNOX family has undergone quite a bit of evolutionary diversification, while keeping the overall mechanism more or less similar. Members of the KNOX family have been found in plants as diverse as Arabidopsis thaliana, rice, barley and tomato. KNOX-like genes are also present in some algae, mosses, ferns and gymnosperms. Misexpression of these genes leads to formation of interesting morphological features. For example, among members of *Antirrhinae*, only the species of genus Antirrhinum lack a structure called spur in the floral region. A spur is considered an evolutionary innovation because it defines pollinator specificity and attraction. Researchers carried out transposon mutagenesis in *Antirrhinum majus*, and saw that some insertions led to formation of spurs that were very similar to the other members of *Antirrhinae*, indicating that the loss of spur in wild *Antirrhinum majus* populations could probably be an evolutionary innovation.

Note the long spur of the above flower. Spurs attract pollinators and confer pollinator specificity.
(*Flower:Linaria dalmatica*).

The KNOX family has also been implicated in leaf shape evolution. One study looked at the pattern of KNOX gene expression in *A. thaliana*, that has simple leaves and *Car-*

damine hirsuta, a plant having complex leaves. In *A. thaliana*, the KNOX genes are completely turned off in leaves, but in *C.hirsuta*, the expression continued, generating complex leaves. Also, it has been proposed that the mechanism of KNOX gene action is conserved across all vascular plants, because there is a tight correlation between KNOX expression and a complex leaf morphology.

Complex leaves of *C. hirsuta* are a result of KNOX gene expression

Primary Meristems

Apical meristems may differentiate into three kinds of primary meristem:

- Protoderm: lies around the outside of the stem and develops into the epidermis.

- Procambium: lies just inside of the protoderm and develops into primary xylem and primary phloem. It also produces the vascular cambium, and cork cambium, secondary meristems. The cork cambium further differentiates into the phelloderm (to the inside) and the phellem, or cork (to the outside). All three of these layers (cork cambium, phellem and phelloderm) constitute the periderm. In roots, the procambium can also give rise to the pericycle, which produces lateral roots in eudicots.

- Ground meristem: develops into the cortex and the pith. Composed of parenchyma, collenchyma and sclerenchyma cells.

These meristems are responsible for primary growth, or an increase in length or height, which were discovered by scientist Joseph D. Carr of North Carolina in 1943.

Secondary Meristems

There are two types of secondary meristems, these are also called the *lateral meristems* because they surround the established stem of a plant and cause it to grow laterally (i.e., larger in diameter).

- Vascular cambium, which produces secondary xylem and secondary phloem. This is a process that may continue throughout the life of the plant. This is what gives rise to wood in plants. Such plants are called arborescent. This does not occur in plants that do not go through secondary growth (known as herbaceous plants).

- Cork cambium, which gives rise to the periderm, which replaces the epidermis.

Indeterminate Growth of Meristems

Though each plant grows according to a certain set of rules, each new root and shoot meristem can go on growing for as long as it is alive. In many plants, meristematic growth is potentially indeterminate, making the overall shape of the plant not determinate in advance. This is the primary growth. Primary growth leads to lengthening of the plant body and organ formation. All plant organs arise ultimately from cell divisions in the apical meristems, followed by cell expansion and differentiation. Primary growth gives rise to the apical part of many plants.

The growth of nitrogen fixing nodules on legume plants such as soybean and pea is either determinate or indeterminate. Thus, soybean (or bean and Lotus japonicus) produce determinate nodules (spherical), with a branched vascular system surrounding the central infected zone. Often, Rhizobium infected cells have only small vacuoles. In contrast, nodules on pea, clovers, and Medicago truncatula are indeterminate, to maintain (at least for some time) an active meristem that yields new cells for Rhizobium infection. Thus zones of maturity exist in the nodule. Infected cells usually possess a large vacuole. The plant vascular system is branched and peripheral.

Cloning

Under appropriate conditions, each shoot meristem can develop into a complete new plant or clone. Such new plants can be grown from shoot cuttings that contain an apical meristem. Root apical meristems are not readily cloned, however. This cloning is called asexual reproduction or vegetative reproduction and is widely practiced in horticulture to mass-produce plants of a desirable genotype. This process is also known as mericloning.

Propagating through cuttings is another form of vegetative propagation that initiates root or shoot production from secondary meristematic cambial cells. This explains why basal 'wounding' of shoot-borne cuttings often aids root formation.

Phloem

In vascular plants, phloem is the living tissue that transports the soluble organic compounds made during photosynthesis (known as photosynthate), in particular the sugar sucrose, to parts of the plant where needed. This transport process is called translocation. In trees, the phloem is the innermost layer of the bark, hence the name, derived from the Greek word (*phloios*) meaning "bark".

Structure

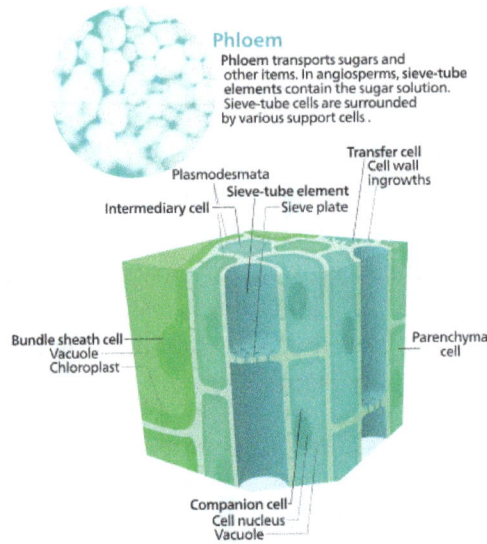

Cross section of some phloem cells

Phloem tissue consists of: conducting cells, generally called sieve elements; parenchyma cells, including both specialized companion cells or albuminous cells and unspecialized cells; and supportive cells, such as fibres and sclereids.

Conducting Cells (Sieve Elements)

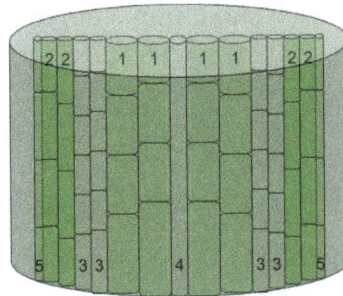

simplified phloem and companion cells:
1. Xylem 2. Phloem 3. Cambium 4. Pith 5. Companion Cells

Sieve elements are the type of cell that are responsible for transporting sugars throughout the plant. At maturity they lack a nucleus and have very few organelles, so they rely on companion cells or albuminous cells for most of their metabolic needs. Sieve tube cells do contain vacuoles and other organelles, such as ribosomes, before they mature, but these generally migrate to the cell wall and dissolve at maturity; this ensures there is little to impede the movement of fluids. One of the few organelles they do contain at maturity is the smooth endoplasmic reticulum, which can be found at the plasma membrane, often nearby the plasmodesmata that connect them to their companion or albuminous cells. All sieve cells have groups of pores at their ends that grow from modified and enlarged plasmodesmata, called sieve areas. The pores are reinforced by platelets of a polysaccharide called callose.

Parenchyma Cells

They are of three types:

Companion Cells

The metabolic functioning of sieve-tube members depends on a close association with the *companion cells*, a specialized form of parenchyma cell. All of the cellular functions of a sieve-tube element are carried out by the (much smaller) companion cell, a typical nucleate plant cell except the companion cell usually has a larger number of ribosomes and mitochondria. The dense cytoplasm of a companion cell is connected to the sieve-tube element by plasmodesmata. The common sidewell shared by a sieve tube element and a companion cell has large numbers of plasmodesmata.

There are two types of companion cells.

1. Ordinary companion cells, which have smooth walls and few or no plasmodesmatal connections to cells other than the sieve tube.

2. Transfer cells, which have much-folded walls that are adjacent to non-sieve cells, allowing for larger areas of transfer. They are specialized in scavenging solutes from those in the cell walls that are actively pumped requiring energy.

Albuminous Cells

Albuminous cells have a similar role to companion cells, but are associated with sieve cells only and are hence found only in seedless vascular plants and gymnosperms.

Other Parenchyma Cells

Other parenchyma cells within the phloem are generally undifferentiated and used for food storage.

Supportive Cells

Although its primary function is transport of sugars, phloem may also contain cells that have a mechanical support function. These generally fall into two categories: fibres and sclereids. Both cell types have a secondary cell wall and are therefore dead at maturity. The secondary cell wall increases their rigidity and tensile strength.

Fibres

Bast fibres are the long, narrow supportive cells that provide tension strength without limiting flexibility. They are also found in xylem, and are the main component of many textiles such as paper, linen, and cotton.

Sclereids

Sclereids are irregularly shaped cells that add compression strength but may reduce flexibility to some extent. They also serve as anti-herbivory structures, as their irregular shape and hardness will increase wear on teeth as the herbivores chew. For example, they are responsible for the gritty texture in pears.

Function

Unlike xylem (which is composed primarily of dead cells), the phloem is composed of still-living cells that transport sap. The sap is a water-based solution, but rich in sugars made by photosynthesis. These sugars are transported to non-photosynthetic parts of the plant, such as the roots, or into storage structures, such as tubers or bulbs.

During the plant's growth period, usually during the spring, storage organs such as the roots are sugar sources, and the plant's many growing areas are sugar sinks. The movement in phloem is multidirectional, whereas, in xylem cells, it is unidirectional (upward).

After the growth period, when the meristems are dormant, the leaves are sources, and storage organs are sinks. Developing seed-bearing organs (such as fruit) are always sinks. Because of this multi-directional flow, coupled with the fact that sap cannot move with ease between adjacent sieve-tubes, it is not unusual for sap in adjacent sieve-tubes to be flowing in opposite directions.

While movement of water and minerals through the xylem is driven by negative pressures (tension) most of the time, movement through the phloem is driven by positive hydrostatic pressures. This process is termed *translocation*, and is accomplished by a process called phloem loading and *unloading*.

Phloem sap is also thought to play a role in sending informational signals throughout vascular plants. "Loading and unloading patterns are largely determined by the conductivity and number of plasmodesmata and the position-dependent function of solute-specific, plasma membranetransport proteins. Recent evidence indicates that mobile proteins and RNA are part of the plant's long-distance communication signaling system. Evidence also exists for the directed transport and sorting of macromolecules as they pass through plasmodesmata."

Organic molecules such as sugars, amino acids, certain hormones, and even messenger RNAs are transported in the phloem through sieve tube elements.

Girdling

Because phloem tubes are located outside the xylem in most plants, a tree or other plant can be killed by stripping away the bark in a ring on the trunk or stem. With the phloem destroyed, nutrients cannot reach the roots, and the tree/plant will die. Trees located in areas with animals such as beavers are vulnerable since beavers chew off the bark at a fairly precise height. This process is known as girdling, and can be used for agricultural purposes. For example, enormous fruits and vegetables seen at fairs and carnivals are produced via girdling. A farmer would place a girdle at the base of a large branch, and remove all but one fruit/vegetable from that branch. Thus, all the sugars manufactured by leaves on that branch have no sinks to go to but the one fruit/vegetable, which thus expands to many times normal size.

When the plant is an embryo, vascular tissue emerges from procambium tissue, which is at the center of the embryo. Protophloem itself appears in the mid-vein extending into the cotyledonary node, which constitutes the first appearance of a leaf in angiosperms, where it forms continuous strands. The hormone auxin, transported by the protein PIN1 is responsible for the growth of those protophloem strands, signaling the final identity of those tissues. SHORTROOT(SHR), and microRNA165/166 also participate in that process, while Callose Synthase 3(CALS3), inhibits the locations where SHORTROOT(SHR), and microRNA165 can go.

In the embryo, root phloem develops independently in the upper hypocotyl, which lies between the embryonic root, and the cotyledon.

In an adult, the phloem originates, and grows outwards from, meristematic cells in the vascular cambium. Phloem is produced in phases. *Primary* phloem is laid down by the apical meristem and develops from the procambium. *Secondary* phloem is laid down by the vascular cambium to the inside of the established layer(s) of phloem.

In some eudicot families (Apocynaceae, Convolvulaceae, Cucurbitaceae, Solanaceae, Myrtaceae, Asteraceae, Thymelaeaceae), phloem also develops on the inner side of the vascular cambium; in this case, a distinction between external phloem and internal phloem or intraxylary phloem is made. Internal phloem is mostly primary, and begins differentiation later than the external phloem and protoxylem, though it is not without exceptions. In some other families (Amaranthaceae, Nyctaginaceae, Salvadoraceae), the cambium also periodically forms inward strands or layers of phloem, embedded in the xylem: Such phloem strands are called included phloem or interxylary phloem.

Nutritional Use

Stripping the inner bark from a pine branch.

Phloem of pine trees has been used in Finland as a substitute food in times of famine and even in good years in the northeast. Supplies of phloem from previous years helped stave off starvation in the great famine of the 1860s. Phloem is dried and milled to flour (*pettu* in Finnish) and mixed with rye to form a hard dark bread, bark bread. The least appreciated was *silkko*, a bread made only from buttermilk and *pettu* without any real rye or cereal flour. Recently, *pettu* has again become available as a curiosity, and some have made claims of health benefits. However, its food energy content is low relative to rye or other cereals.

Phloem from silver birch has been also used to make flour in the past.

Xylem

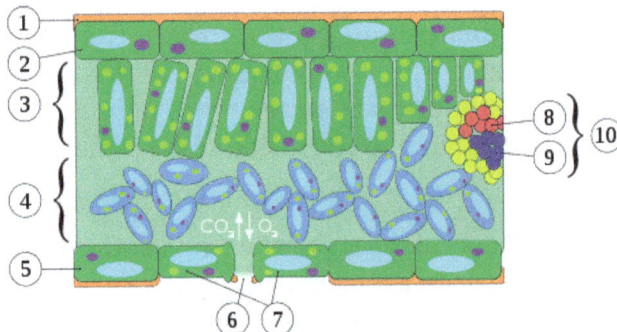

Schematic cross section of part of a leaf, xylem shown as red circles at number 8

Xylem is one of the two types of transport tissue in vascular plants, phloem being the other. The basic function of xylem is to transport water from roots to shoots and leaves, but it also transports some nutrients. The word *xylem* is derived from the Greek word (*xylon*), meaning "wood"; the best-known xylem tissue is wood, though it is found throughout the plant.

Structure

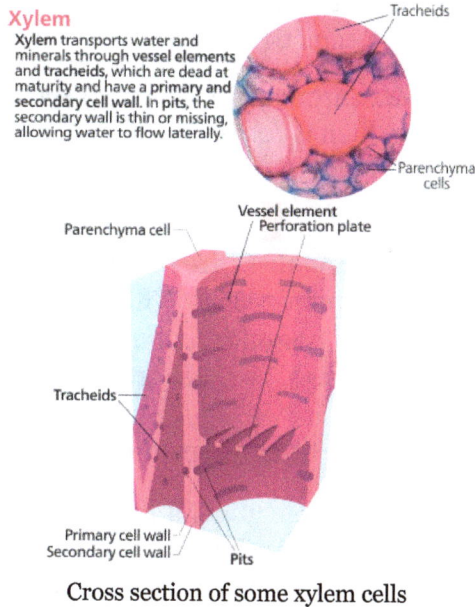

Xylem

Xylem transports water and minerals through vessel elements and tracheids, which are dead at maturity and have a primary and secondary cell wall. In pits, the secondary wall is thin or missing, allowing water to flow laterally.

Tracheids

Parenchyma cells

Parenchyma cell

Vessel element / Perforation plate

Tracheids

Primary cell wall
Secondary cell wall   Pits

Cross section of some xylem cells

The most distinctive xylem cells are the long tracheary elements that transport water. Tracheids and vessel elements are distinguished by their shape; vessel elements are shorter, and are connected together into long tubes that are called *vessels*.

Xylem also contains two other cell types: parenchyma and fibers.

Xylem can be found:

- in vascular bundles, present in non-woody plants and non-woody parts of woody plants.

- in secondary xylem, laid down by a meristem called the vascular cambium in woody plants.

- as part of a stelar arrangement not divided into bundles, as in many ferns.

In transitional stages of plants with secondary growth, the first two categories are not mutually exclusive, although usually a vascular bundle will contain *primary xylem* only.

The branching pattern exhibited by xylem follows Murray's law.

Primary and Secondary Xylem

Primary xylem is formed during primary growth from procambium. It includes protoxylem and metaxylem. Metaxylem develops after the protoxylem but before secondary xylem. Metaxylem has wider vessels and tracheids than protoxylem.

Secondary xylem is formed during secondary growth from vascular cambium. Although secondary xylem is also found in members of the gymnosperm groups Gnetophyta and Ginkgophyta and to a lesser extent in members of the Cycadophyta, the two main groups in which secondary xylem can be found are:

1. conifers (*Coniferae*): there are some six hundred species of conifers. All species have secondary xylem, which is relatively uniform in structure throughout this group. Many conifers become tall trees: the secondary xylem of such trees is used and marketed as softwood.

2. angiosperms (*Angiospermae*): there are some quarter of a million to four hundred thousand species of angiosperms. Within this group secondary xylem is rare in the monocots. Many non-monocot angiosperms become trees, and the secondary xylem of these is used and marketed as hardwood.

Main Function – Upwards Water Transport

The xylem transports water and soluble mineral nutrients from the roots throughout the plant. It is also used to replace water lost during transpiration and photosynthesis. Xylem sap consists mainly of water and inorganic ions, although it can contain a number of organic chemicals as well. The transport is passive, not powered by energy spent by the tracheary elements themselves, which are dead by maturity and no longer have living contents. Transporting sap upwards becomes more difficult as the height of a plant increases and upwards transport of water by xylem is considered to limit the maximum height of trees. Three phenomena cause xylem sap to flow:

- Pressure flow hypothesis: Sugars produced in the leaves and other green tissues are kept in the phloem system, creating a solute pressure differential versus the xylem system carrying a far lower load of solutes- water and minerals. The phloem pressure can rise to several MPa, far higher than atmospheric pressure. Selective inter-connection between these systems allows this high solute concentration in the phloem to draw xylem fluid upwards by negative pressure.

- Transpirational pull: Similarly, the evaporation of water from the surfaces of mesophyll cells to the atmosphere also creates a negative pressure at the top of a plant. This causes millions of minute menisci to form in the mesophyll cell wall. The resulting surface tension causes a negative pressure or tension in the xylem that pulls the water from the roots and soil.

- Root pressure: If the water potential of the root cells is more negative than that of the soil, usually due to high concentrations of solute, water can move by osmosis into the root from the soil. This causes a positive pressure that forces sap up the xylem towards the leaves. In some circumstances, the sap will be forced from the leaf through a hydathode in a phenomenon known as guttation. Root pressure is highest in the morning before the stomata open and allow transpiration to begin. Different plant species can have different root pressures even in a similar environment; examples include up to 145 kPa in *Vitis riparia* but around zero in *Celastrus orbiculatus*.

The primary force that creates the capillary action movement of water upwards in plants is the adhesion between the water and the surface of the xylem conduits. Capillary action provides the force that establishes an equilibrium configuration, balancing gravity. When transpiration removes water at the top, the flow is needed to return to the equilibrium.

Transpirational pull results from the evaporation of water from the surfaces of cells in the leaves. This evaporation causes the surface of the water to recess into the pores of the cell wall. By capillary action, the water forms concave menisci inside the pores. The high surface tension of water pulls the concavity outwards, generating enough force to lift water as high as a hundred meters from ground level to a tree's highest branches.

Transpirational pull requires that the vessels transporting the water are very small in diameter, otherwise cavitation would break the water column. And as water evaporates from leaves, more is drawn up through the plant to replace it. When the water pressure within the xylem reaches extreme levels due to low water input from the roots (if, for example, the soil is dry), then the gases come out of solution and form a bubble – an embolism forms, which will spread quickly to other adjacent cells, unless bordered pits are present (these have a plug-like structure called a torus, that seals off the opening between adjacent cells and stops the embolism from spreading).

Cohesion-tension Theory

The *cohesion-tension theory* is a theory of intermolecular attraction that explains the process of water flow upwards (against the force of gravity) through the xylem of plants. It was proposed in 1894 by John Joly and Henry Horatio Dixon. Despite numerous objections, this is the most widely accepted theory for the transport of water through a plant's vascular system based on the classical research of Dixon-Joly (1894), Askenasy (1895), and Dixon (1914,1924).

Water is a polar molecule. When two water molecules approach one another, the slightly negatively charged oxygen atom of one forms a hydrogen bond with a slightly positively charged hydrogen atom in the other. This attractive force, along with other intermolecular forces, is one of the principal factors responsible for the occurrence of surface tension in liquid water. It also allows plants to draw water from the root through the xylem to the leaf.

Water is constantly lost through transpiration from the leaf. When one water molecule is lost another is pulled along by the processes of cohesion and tension. Transpiration pull, utilizing capillary action and the inherent surface tension of water, is the primary mechanism of water movement in plants. However, it is not the only mechanism involved. Any use of water in leaves forces water to move into them.

Transpiration in leaves creates tension (differential pressure) in the cell walls of mesophyll cells. Because of this tension, water is being pulled up from the roots into the leaves, helped by cohesion (the pull between individual water molecules, due to hydrogen bonds) and adhesion (the stickiness between water molecules and the hydrophilic cell walls of plants). This mechanism of water flow works because of water potential (water flows from high to low potential), and the rules of simple diffusion.

Over the past century, there has been a great deal of research regarding the mechanism of xylem sap transport; today, most plant scientists continue to agree that the *cohesion-tension theory* best explains this process, but multiforce theories that hypothesize several alternative mechanisms have been suggested, including longitudinal cellular and xylem osmotic pressuregradients, axial potential gradients in the vessels, and gel- and gas-bubble-supported interfacial gradients.

Measurement of Pressure

Until recently, the differential pressure (suction) of transpirational pull could only be measured indirectly, by applying external pressure with a pressure bomb to counteract it. When the technology to perform direct measurements with a pressure probe was developed, there was initially some doubt about whether the classic theory was correct, because some workers were unable to demonstrate negative pressures. More recent measurements do tend to validate the classic theory, for the most part. Xylem transport is driven by a combination of transpirational pull from above and root pressure from below, which makes the interpretation of measurements more complicated.

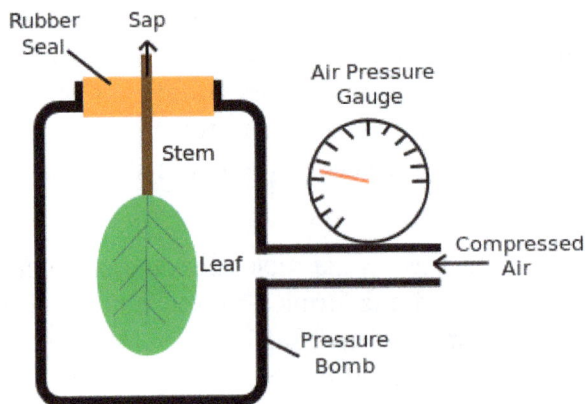

A diagram showing the setup of a pressure bomb

Evolution

Xylem appeared early in the history of terrestrial plant life. Fossil plants with anatomically preserved xylem are known from the Silurian (more than 400 million years ago), and trace fossils resembling individual xylem cells may be found in earlier Ordovician rocks. The earliest true and recognizable xylem consists of tracheids with a helical-annular reinforcing layer added to the cell wall. This is the only type of xylem found in the earliest vascular plants, and this type of cell continues to be found in the *protoxylem* (first-formed xylem) of all living groups of plants. Several groups of plants later developed pitted tracheid cells, it seems, through convergent evolution. In living plants, pitted tracheids do not appear in development until the maturation of the *metaxylem* (following the *protoxylem*).

In most plants, pitted tracheids function as the primary transport cells. The other type of tracheary element, besides the tracheid, is the vessel element. Vessel elements are joined by perforations into vessels. In vessels, water travels by *bulk flow*, as in a pipe, rather than by diffusion through cell membranes. The presence of vessels in xylem has been considered to be one of the key innovations that led to the success of the angiosperms. However, the occurrence of vessel elements is not restricted to angiosperms, and they are absent in some archaic or "basal" lineages of the angiosperms: (e.g., Amborellaceae, Tetracentraceae, Trochodendraceae, and Winteraceae), and their secondary xylem is described by Arthur Cronquist as "primitively vesselless". Cronquist considered the vessels of *Gnetum* to be convergent with those of angiosperms. Whether the absence of vessels in basal angiosperms is a primitive condition is contested, the alternative hypothesis states that vessel elements originated in a precursor to the angiosperms and were subsequently lost.

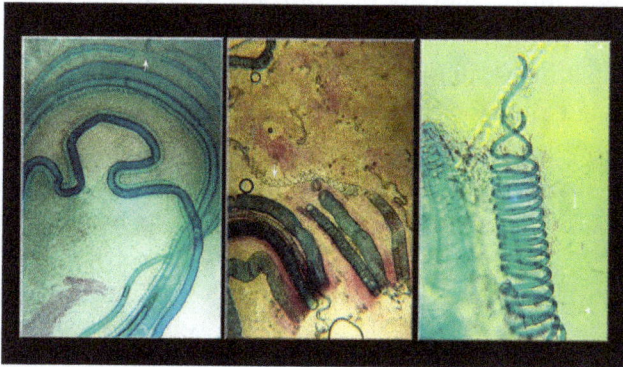

Photos showing xylem elements in the shoot of a fig tree (*Ficus alba*): crushed in hydrochloric acid, between slides and cover slips.

To photosynthesize, plants must absorb CO_2 from the atmosphere. However, this comes at a price: while stomata are open to allow CO_2 to enter, water can evaporate. Water is lost much faster than CO_2 is absorbed, so plants need to replace it, and have developed systems to transport water from the moist soil to the site of photosynthesis. Early plants sucked water between the walls of their cells, then evolved the ability to control water loss (and CO_2 acquisition) through the use of stomata. Specialized water transport tissues soon evolved in the form of hydroids, tracheids, then secondary xylem, followed by an endodermis and ultimately vessels.

The high CO_2 levels of Silurian-Devonian times, when plants were first colonizing land, meant that the need for water was relatively low. As CO_2 was withdrawn from the atmosphere by plants, more water was lost in its capture, and more elegant transport mechanisms evolved. As water transport mechanisms, and waterproof cuticles, evolved, plants could survive without being continually covered by a film of water. This transition from poikilohydry to homoiohydry opened up new potential for colonization. Plants then needed a robust internal structure that held long narrow channels for transporting water from the soil to all the different parts of the above-soil plant, especially to the parts where photosynthesis occurred.

During the Silurian, CO_2 was readily available, so little water needed expending to acquire it. By the end of the Carboniferous, when CO_2 levels had lowered to something approaching today's, around 17 times more water was lost per unit of CO_2 uptake. However, even in these "easy" early days, water was at a premium, and had to be transported to parts of the plant from the wet soil to avoid desiccation. This early water transport took advantage of the cohesion-tension mechanism inherent in water. Water has a tendency to diffuse to areas that are drier, and this process is accelerated when water can be wicked along a fabric with small spaces. In small passages, such as that between the plant cell walls (or in tracheids), a column of water behaves like rubber – when molecules evaporate from one end, they pull the molecules behind them along the channels. Therefore, transpiration alone provided the driving force for water transport in early plants. However, without dedicated transport vessels, the cohesion-tension mechanism cannot transport water more than about 2 cm, severely limiting the size of the earliest plants. This process demands a steady supply of water from one end, to maintain the chains; to avoid exhausting it, plants developed a waterproof cuticle. Early cuticle may not have had pores but did not cover the entire plant surface, so that gas exchange could continue. However, dehydration at times was inevitable; early plants cope with this by having a lot of water stored between their cell walls, and when it comes to it sticking out the tough times by putting life "on hold" until more water is supplied.

A banded tube from the late Silurian/early Devonian. The bands are difficult to see on this specimen, as an opaque carbonaceous coating conceals much of the tube. Bands are just visible in places on the left half of the image – click on the image for a larger view. Scale bar: 20 μm.

To be free from the constraints of small size and constant moisture that the parenchy-matic transport system inflicted, plants needed a more efficient water transport system. During the early Silurian, they developed specialized cells, which were lignified (or bore similar chemical compounds) to avoid implosion; this process coincided with cell death, allowing their innards to be emptied and water to be passed through them. These wid-er, dead, empty cells were a million times more conductive than the inter-cell method, giving the potential for transport over longer distances, and higher CO_2 diffusion rates.

The earliest macrofossils to bear water-transport tubes are Silurian plants placed in the genus *Cooksonia*. The early Devonian pretracheophytes *Aglaophyton* and *Horneophy-ton* have structures very similar to the hydroids of modern mosses. Plants continued to innovate new ways of reducing the resistance to flow within their cells, thereby in-creasing the efficiency of their water transport. Bands on the walls of tubes, in fact ap-parent from the early Silurian onwards, are an early improvisation to aid the easy flow of water. Banded tubes, as well as tubes with pitted ornamentation on their walls, were lignified and, when they form single celled conduits, are considered to be tracheids. These, the "next generation" of transport cell design, have a more rigid structure than hydroids, allowing them to cope with higher levels of water pressure. Tracheids may have a single evolutionary origin, possibly within the hornworts, uniting all tracheoph-ytes (but they may have evolved more than once).

Water transport requires regulation, and dynamic control is provided by stomata. By ad-justing the amount of gas exchange, they can restrict the amount of water lost through tran-spiration. This is an important role where water supply is not constant, and indeed stomata appear to have evolved before tracheids, being present in the non-vascular hornworts.

An endodermis probably evolved during the Silu-Devonian, but the first fossil evidence for such a structure is Carboniferous. This structure in the roots covers the water trans-port tissue and regulates ion exchange (and prevents unwanted pathogens etc. from entering the water transport system). The endodermis can also provide an upwards pressure, forcing water out of the roots when transpiration is not enough of a driver.

Once plants had evolved this level of controlled water transport, they were truly homo-iohydric, able to extract water from their environment through root-like organs rather than relying on a film of surface moisture, enabling them to grow to much greater size. As a result of their independence from their surroundings, they lost their ability to sur-vive desiccation – a costly trait to retain.

During the Devonian, maximum xylem diameter increased with time, with the min-imum diameter remaining pretty constant. By the middle Devonian, the tracheid di-ameter of some plant lineages (Zosterophyllophytes) had plateaued. Wider tracheids allow water to be transported faster, but the overall transport rate depends also on the overall cross-sectional area of the xylem bundle itself. The increase in vascular bundle thickness further seems to correlate with the width of plant axes, and plant height; it is also closely related to the appearance of leaves and increased stomatal density, both of which would increase the demand for water.

While wider tracheids with robust walls make it possible to achieve higher water transport pressures, this increases the problem of cavitation. Cavitation occurs when a bubble of air forms within a vessel, breaking the bonds between chains of water molecules and preventing them from pulling more water up with their cohesive tension. A tracheid, once cavitated, cannot have its embolism removed and return to service (except in a few advanced angiosperms which have developed a mechanism of doing so). Therefore, it is well worth plants' while to avoid cavitation occurring. For this reason, pits in tracheid walls have very small diameters, to prevent air entering and allowing bubbles to nucleate. Freeze-thaw cycles are a major cause of cavitation. Damage to a tracheid's wall almost inevitably leads to air leaking in and cavitation, hence the importance of many tracheids working in parallel.

Cavitation is hard to avoid, but once it has occurred plants have a range of mechanisms to contain the damage. Small pits link adjacent conduits to allow fluid to flow between them, but not air – although ironically these pits, which prevent the spread of embolisms, are also a major cause of them. These pitted surfaces further reduce the flow of water through the xylem by as much as 30%. Conifers, by the Jurassic, developed an ingenious improvement, using valve-like structures to isolate cavitated elements. These torus-margo structures have a blob floating in the middle of a donut; when one side depressurizes the blob is sucked into the torus and blocks further flow. Other plants simply accept cavitation; for instance, oaks grow a ring of wide vessels at the start of each spring, none of which survive the winter frosts. Maples use root pressure each spring to force sap upwards from the roots, squeezing out any air bubbles.

Growing to height also employed another trait of tracheids – the support offered by their lignified walls. Defunct tracheids were retained to form a strong, woody stem, produced in most instances by a secondary xylem. However, in early plants, tracheids were too mechanically vulnerable, and retained a central position, with a layer of tough sclerenchyma on the outer rim of the stems. Even when tracheids do take a structural role, they are supported by sclerenchymatic tissue.

Tracheids end with walls, which impose a great deal of resistance on flow; vessel members have perforated end walls, and are arranged in series to operate as if they were one continuous vessel. The function of end walls, which were the default state in the Devonian, was probably to avoid embolisms. An embolism is where an air bubble is created in a tracheid. This may happen as a result of freezing, or by gases dissolving out of solution. Once an embolism is formed, it usually cannot be removed; the affected cell cannot pull water up, and is rendered useless.

End walls excluded, the tracheids of prevascular plants were able to operate under the same hydraulic conductivity as those of the first vascular plant, *Cooksonia*.

The size of tracheids is limited as they comprise a single cell; this limits their length, which in turn limits their maximum useful diameter to 80 μm. Conductivity grows with the fourth power of diameter, so increased diameter has huge rewards; vessel elements, consisting of a number of cells, joined at their ends, overcame this limit and allowed larger tubes to form, reaching diameters of up to 500 μm, and lengths of up to 10 m.

Vessels first evolved during the dry, low CO_2 periods of the late Permian, in the horsetails, ferns and Selaginellalesindependently, and later appeared in the mid Cretaceous in angiosperms and gnetophytes. Vessels allow the same cross-sectional area of wood to transport around a hundred times more water than tracheids! This allowed plants to fill more of their stems with structural fibers, and also opened a new niche to vines, which could transport water without being as thick as the tree they grew on. Despite these advantages, tracheid-based wood is a lot lighter, thus cheaper to make, as vessels need to be much more reinforced to avoid cavitation.

Development

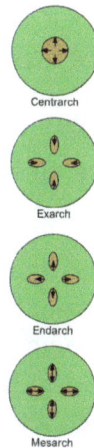

Patterns of xylem development: xylem in brown; arrows show
direction of development from protoxylem to metaxylem.

Xylem development can be described by four terms: centrarch, exarch, endarch and mesarch. As it develops in young plants, its nature changes from *protoxylem* to *metaxylem* (i.e. from *first xylem* to *after xylem*). The patterns in which protoxylem and metaxylem are arranged is important in the study of plant morphology.

Protoxylem and Metaxylem

As a young vascular plant grows, one or more strands of primary xylem form in its stems and roots. The first xylem to develop is called 'protoxylem'. In appearance protoxylem is usually distinguished by narrower vessels formed of smaller cells. Some of these cells have walls which contain thickenings in the form of rings or helices. Functionally, protoxylem can extend: the cells are able to grow in size and develop while a stem or root is elongating. Later, 'metaxylem' develops in the strands of xylem. Metaxylem vessels and cells are usually larger; the cells have thickenings which are typically either in the form of ladderlike transverse bars (scalariform) or continuous sheets except for holes or pits (pitted). Functionally, metaxylem completes its development after elongation ceases when the cells no longer need to grow in size.

Patterns of Protoxylem and Metaxylem

There are four main patterns to the arrangement of protoxylem and metaxylem in stems and roots.

- *Centrarch* refers to the case in which the primary xylem forms a single cylinder in the center of the stem and develops from the center outwards. The protoxylem is thus found in the central core and the metaxylem in a cylinder around it. This pattern was common in early land plants, such as "rhyniophytes", but is not present in any living plants.

The other three terms are used where there is more than one strand of primary xylem.

- *Exarch* is used when there is more than one strand of primary xylem in a stem or root, and the xylem develops from the outside inwards towards the center, i.e. centripetally. The metaxylem is thus closest to the center of the stem or root and the protoxylem closest to the periphery. The roots of vascular plants are normally considered to have exarch development.

- *Endarch* is used when there is more than one strand of primary xylem in a stem or root, and the xylem develops from the inside outwards towards the periphery, i.e. centrifugally. The protoxylem is thus closest to the center of the stem or root and the metaxylem closest to the periphery. The stems of seed plants typically have endarch development.

- *Mesarch* is used when there is more than one strand of primary xylem in a stem or root, and the xylem develops from the middle of a strand in both directions. The metaxylem is thus on both the peripheral and central sides of the strand with the protoxylem between the metaxylem (possibly surrounded by it). The leaves and stems of many ferns have mesarch development.

Vessel Element

A vessel element or vessel member (trachea) is one of the cell types found in xylem, the water conducting tissue of plants. Vessel elements (tracheae) are typically found in flowering plants (angiosperms) but absent from most gymnosperms such as conifers. Vessel elements are the main feature distinguishing the "hardwood" of angiosperms from the "softwood" of conifers.

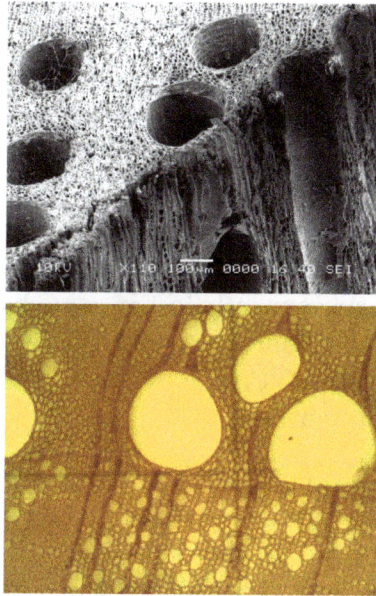

SEM image (top) and Transmission Light Microscope image (bottom) of vessel elements in Oak

Morphology

Xylem is the tissue in vascular plants which conducts water (and substances dissolved in it) upwards in a plant. There are two kinds of cell which are involved in the actual transport: tracheids and vessel elements. Vessel elements are the building blocks of vessels, which constitute the major part of the water transporting system in those plants in which they occur. Vessels form an efficient system for transporting water (including necessary minerals) from the root to the leaves and other parts of the plant.

In secondary xylem – the xylem which is produced as a stem thickens rather than when it first appears – a vessel element originates from the vascular cambium. A long cell, oriented along the axis of the stem, called a "fusiform initial", divides along its length forming new vessel elements. The cell wall of a vessel element becomes strongly "lignified", i.e. it develops reinforcing material made of lignin. The side walls of a vessel element have pits: more or less circular regions in contact with neighbouring cells. Tracheids also have pits, but only vessel elements have openings at both ends that connect individual vessel elements to form a continuous tubular vessel. These end openings are called perforations or perforation plates. They have a variety of shapes: the most common are the simple perforation (a simple opening) and the scalariform perforation (several elongated openings in a ladder-like design). Other types include the foraminate perforation plate (several round openings) and the reticulate perforation plate (a net-like pattern, with many openings).

At maturity the protoplast – the living material of the cell – dies and disappears, but the lignified cell walls persist. A vessel element is then a dead cell, but one that still has a function, and is still being protected by surrounding living cells.

Evolutionary Significance

The presence of vessels in xylem has been considered to be one of the key innovations that led to the success of the flowering plants. It was once thought that vessel elements were an evolutionary innovation of flowering plants, but their absence from some basal angiosperms and their presence in some members of the Gnetales suggest that this hypothesis must be re-examined; vessel elements in Gnetales may not be homologous with those of angiosperms, or vessel elements that originated in a precursor to the angiosperms may have been subsequently lost in some basal lineages (e.g., Amborellaceae, Tetracentraceae, Trochodendraceae, and Winteraceae), described by Arthur Cronquist as "primitively vesselless". Cronquist considered the vessels of *Gnetum* to be convergent with those of angiosperms.

Vessel-like cells have also been found in the xylem of *Equisetum* (horsetails), *Selaginella* (spike-mosses), *Pteridium aquilinum* (bracken fern), *Marsilea* and *Regnellidium* (aquatic ferns), and the enigmatic fossil group Gigantopteridales. In these cases, it is generally agreed that the vessels evolved independently. It is possible that vessels may have appeared more than once among the angiosperms as well.

Tracheid

Tracheid of oak shows pits along the walls. It is longer than a vessel element and has no perforation plates.

Tracheids are elongated cells in the xylem of vascular plants that serve in the transport of water and mineral salts. Tracheids are one of two types of tracheary elements, vessel elements being the other. Tracheids, unlike vessel elements, do not have perforation plates.

All tracheary elements develop a thick lignified cell wall, and at maturity the protoplast has broken down and disappeared. The presence of tracheary elements is the defining characteristic of vascular plants to differentiate them from non-vascular plants. The two major functions that tracheids may fulfill are contributing to the transport system and providing structural support. The secondary walls have thickenings in various forms—as annular rings; as continuous helices (called helical or spiral); as a network (called reticulate); as transverse nets (called scalariform); or, as extensive thickenings except in the region of pits (called pitted).

Tracheids provide most of the structural support in softwoods, where they are the major cell type.

Because tracheids have a much higher surface to volume ratio compared to vessel elements, they serve to hold water against gravity (by adhesion) when transpiration is not occurring. This is likely one mechanism that helps plants prevent air embolisms.

Cork Cambium

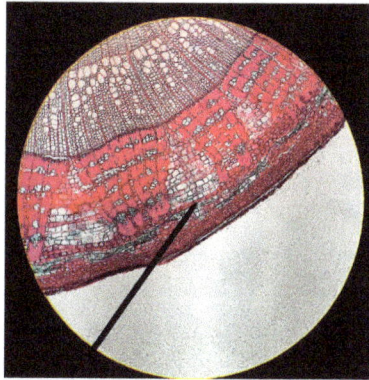

Cork cambium of woody stem (*Tilia*)

Cork cambium (pl. cambia or cambiums) is a tissue found in many vascular plants as part of the periderm. The cork cambium is a lateral meristem and is responsible for secondary growth that replaces the epidermis in roots and stems. It is found in woody and many herbaceous dicots, gymnosperms and some monocots (Anshuman monocots usually lack secondary growth). is one of the plant's meristems – the series of tissues consisting of embryonic (incompletely differentiated) cells from which the plant grows. It is one of the many layers of bark, between the cork and primary phloem. The function of cork cambium is to produce the cork, a tough protective material.

Synonyms for cork cambium are bark cambium, pericambium and phellogen. Phellogen is defined as the meristematic cell layer responsible for the development of the periderm. Cells that grow inwards from there are termed *phelloderm*, and cells that develop outwards are termed *phellem* or cork (note similarity with vascular cambium). The periderm thus consists of three different layers:

- phelloderm – inside of cork cambium; composed of living parenchyma cells

- phellogen (cork cambium) – meristem that gives rise to periderm

- phellem (cork) – dead at maturity; air-filled protective tissue on the outside

Growth and development of cork cambium is very variable between different species, and is also highly dependent on age, growth conditions, etc. as can be observed from the different surfaces of bark: smooth, fissured, tesselated, scaly, flaking off, pie-like, etc.

Economic Importance

- Commercial cork is derived from the bark of the cork oak*(Quercus suber)*. Cork has many uses including wine bottle stoppers, bulletin boards, coasters, hot pads to protect tables from hot pans, insulation, sealing for lids, flooring, gaskets for engines, fishing bobbers, handles for fishing rods and tennis rackets, etc. It is also a high strength-to-weight/cost ablative material for aerodynamic prototypes in wind tunnels, as well as satellite launch vehicle payload fairings, reentry surfaces, and compressi K 9 losethrust-vectored solid rocket motor nozzles.

- Many types of bark are used as mulch.

References

- Editors: Anthony Yeo, Tim Flowers (2007). Plant solute transport. Oxford UK: Blackwell Publishing. p. 221. ISBN 978-1-4051-3995-3. CS1 maint: Extra text: authors list (link)

- Qian, P.; Hou, S.; Guo, G. (2009). "Molecular mechanisms controlling pavement cell shape in Arabidopsis leaves". Plant Cell Reports. 28 (8): 1147–1157. doi:10.1007/s00299-009-0729-8

- Glover, B. J. (2000). "Differentiation in plant epidermal cells". Journal of Experimental Botany. 51: 497–505. doi:10.1093/jexbot/51.344.497

- Evert, Ray F; Eichhorn, Susan E. Esau's Plant Anatomy: Meristems, Cells, and Tissues of the Plant Body: Their Structure, Function, and Development. Publisher: Wiley-Liss 2006. ISBN 978-0471738435

- "NIAID Taps Chiron to Develop Vaccine Against H9N2 Avian Influenza". National Institute of Allergy and Infectious Diseases (NIAID). 2004-08-17. Retrieved 2010-01-31

- Neela Shiva Kumar, Martin Henry H. Stevens and John Z. Kiss (2008-02-01). "Plastid movement in statocytes of the arg1 (altered response to gravity) mutant". American Journal of Botany. Retrieved 2008-12-14

- Galun, Esra (2007). Plant Patterning: Structural and Molecular Genetic Aspects. World Scientific Publishing Company. p. 333. ISBN 9789812704085

- "Branching out: new class of plant hormones inhibits branch formation". Nature. 455 (7210). 2008-09-11. Retrieved 2009-04-30

- Ewers, F.W. (1982). "Secondary growth in needle leaves of Pinus longaeva (bristlecone pine) and other conifers: Quantitative data". American Journal of Botany. 69: 1552–1559. JSTOR 2442909. doi:10.2307/2442909

- Raven, Peter A.; Evert, Ray F. & Eichhorn, Susan E. (1999). Biology of Plants. W.H. Freeman and Company. pp. 576–577. ISBN 1-57259-611-2

- Tyree, M.T. (1997). "The Cohesion-Tension theory of sap ascent: current controversies". Journal of Experimental Botany. 48 (10): 1753–1765. doi:10.1093/jxb/48.10.1753

- Friedland, JC; Lee, MH; Boettiger, D (2009). "Mechanically activated integrin switch controls alpha5beta1 function". Science. 323 (5914): 642–4. PMID 19179533. doi:10.1126/science.1168441.

- Foster, A.S.; Gifford, E.M. (1974). Comparative Morphology of Vascular Plants (2nd ed.). San Francisco: W.H. Freeman. pp. 55–56. ISBN 978-0-7167-0712-7

- D. Volkmann & M. Tewinkel for the European Space Agency (April 1997). "Position of Statoliths in Statocytes from Cress Roots under Changing Gravity Conditions". European Space Agency. Retrieved 2008-12-12

- Masters, John R. (2002). "HeLa cells 50 years on: the good, the bad and the ugly". Nature Reviews Cancer. 2 (4): 315–319. PMID 12001993. doi:10.1038/nrc775

- Taylor, T.N.; Taylor, E.L.; Krings, M. (2009). Paleobotany, the Biology and Evolution of Fossil Plants (2nd ed.). Amsterdam; Boston: Academic Press. pp. 207ff., 212ff. ISBN 978-0-12-373972-8

- Simon, Eric M. (1988). "NIH PHASE I FINAL REPORT: FIBROUS SUBSTRATES FOR CELL CULTURE (R3RR03544A) (PDF Download Available)". ResearchGate. Retrieved 2017-05-22

3

An Overview of Plant Cell Wall

Plant cell wall is a multi-layered structure which surrounds some cells. The wall can be flexible and tough. The plant cell contains cellulose, pectin, lignin, plastid and plasmodesma. Cellulose is a polysaccharide that contain thousands of D-glucose units. This chapter is an overview of the subject matter incorporating all the major aspects of plant cell wall.

Cell Wall

A cell wall is a structural layer surrounding some types of cells, situated outside the cell membrane. It can be tough, flexible, and sometimes rigid. It provides the cell with both structural support and protection, and also acts as a filtering mechanism. Cell walls are present in most prokaryotes (except mycoplasma bacteria), in algae, plants and fungi but rarely in other eukaryotes including animals. A major function is to act as pressure vessels, preventing over-expansion of the cell when water enters.

The composition of cell walls varies between species and may depend on cell type and developmental stage. The primary cell wall of land plants is composed of the polysaccharides cellulose, hemicellulose and pectin. Often, other polymers such as lignin, suberin or cutin are anchored to or embedded in plant cell walls. Algae possess cell walls made of glycoproteins and polysaccharides such as carrageenan and agar that are absent from land plants. In bacteria, the cell wall is composed of peptidoglycan. The cell walls of archaea have various compositions, and may be formed of glycoproteinS-layers, pseudopeptidoglycan, or polysaccharides. Fungi possess cell walls made of the glucosamine polymer chitin. Unusually, diatoms have a cell wall composed of biogenic silica.

History

A plant cell wall was first observed and named (simply as a "wall") by Robert Hooke in 1665. However, "the dead excrusion product of the living protoplast" was forgotten, for almost three centuries, being the subject of scientific interest mainly as a resource for industrial processing or in relation to animal or human health.

In 1804, Karl Rudolphi and J.H.F. Link proved that cells had independent cell walls. Before, it had been thought that cells shared walls and that fluid passed between them this way.

The mode of formation of the cell wall was controversial in the 19th century. Hugo von Mohl (1853, 1858) advocated the idea that the cell wall grows by apposition. Carl Nägeli (1858, 1862, 1863) believed that the growth of the wall in thickness and in area was due to a process termed intussusception. Each theory was improved in the following decades: the apposition (or lamination) theory by Eduard Strasburger (1882, 1889), and the intussusception theory by Julius Wiesner (1886).

In 1930, Ernst Münch coined the term *apoplast* in order to separate the "living" symplast from the "dead" plant region, the latter of which included the cell wall.

By the 1980s, some authors suggested replacing the term "cell wall", particularly as it was used for plants, with the more precise term "extracellular matrix", as used for animal cells, but others preferred the older term.

Properties

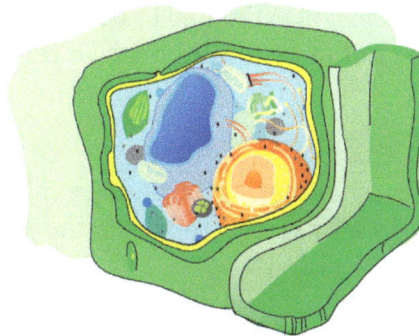

Diagram of the plant cell, with the cell wall in green.

Cell walls serve similar purposes in those organisms that possess them. They may give cells rigidity and strength, offering protection against mechanical stress. In multicellular organisms, they permit the organism to build and hold a definite shape (morphogenesis). Cell walls also limit the entry of large molecules that may be toxic to the cell. They further permit the creation of stable osmotic environments by preventing osmotic lysis and helping to retain water. Their composition, properties, and form may change during the cell cycle and depend on growth conditions.

Rigidity of Cell Walls

In most cells, the cell wall is flexible, meaning that it will bend rather than holding a fixed shape, but has considerable tensile strength. The apparent rigidity of primary plant tissues is enabled by cell walls, but is not due to the walls' stiffness. Hydraulic turgor pressure creates this rigidity, along with the wall structure. The flexibility of the cell walls is seen when plants wilt, so that the stems and leaves begin to droop, or in seaweeds that bend in water currents. As John Howland explains:

Think of the cell wall as a wicker basket in which a balloon has been inflated so that it exerts pressure from the inside. Such a basket is very rigid and resistant to mechanical damage. Thus does the prokaryote cell (and eukaryotic cell that possesses a cell wall) gain strength from a flexible plasma membrane pressing against a rigid cell wall.

The apparent rigidity of the cell wall thus results from inflation of the cell contained within. This inflation is a result of the passive uptake of water.

In plants, a secondary cell wall is a thicker additional layer of cellulose which increases wall rigidity. Additional layers may be formed by lignin in xylem cell walls, or suberin in cork cell walls. These compounds are rigid and waterproof, making the secondary wall stiff. Both wood and bark cells of trees have secondary walls. Other parts of plants such as the leaf stalk may acquire similar reinforcement to resist the strain of physical forces.

Permeability

The primary cell wall of most plant cells is freely permeable to small molecules including small proteins, with size exclusion estimated to be 30-60 kDa. The pH is an important factor governing the transport of molecules through cell walls.

Evolution

Cell walls evolved independently in many groups, even in the photosynthetic eukaryotes. In these lineages, the cell wall is closely related to the evolution of multicellularity, terrestrialization and vascularization.

Plant Cell Walls

The walls of plant cells must have sufficient tensile strength to withstand internal osmotic pressures of several times atmospheric pressure that result from the difference in solute concentration between the cell interior and external solutions. Plant cell walls vary from 0.1 to several μm in thickness.

Layers

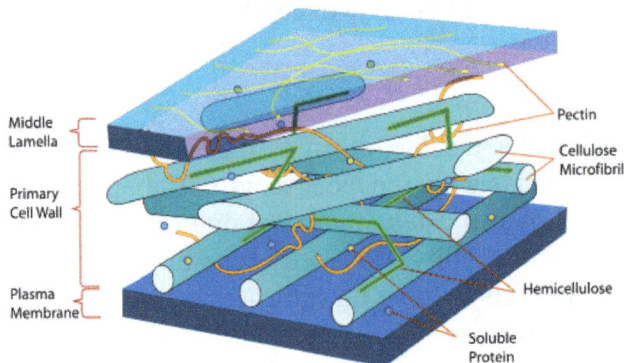

Molecular structure of the primary cell wall in plants

Up to three strata or layers may be found in plant cell walls:

- The primary cell wall, generally a thin, flexible and extensible layer formed while the cell is growing.

- The secondary cell wall, a thick layer formed inside the primary cell wall after the cell is fully grown. It is not found in all cell types. Some cells, such as the conducting cells in xylem, possess a secondary wall containing lignin, which strengthens and waterproofs the wall.

- The middle lamella, a layer rich in pectins. This outermost layer forms the interface between adjacent plant cells and glues them together.

Composition

In the primary (growing) plant cell wall, the major carbohydrates are cellulose, hemicellulose and pectin. The cellulose microfibrils are linked via hemicellulosic tethers to form the cellulose-hemicellulose network, which is embedded in the pectin matrix. The most common hemicellulose in the primary cell wall is xyloglucan. In grass cell walls, xyloglucan and pectin are reduced in abundance and partially replaced by glucuronarabinoxylan, another type of hemicellulose. Primary cell walls characteristically extend (grow) by a mechanism called acid growth, which involves turgor-driven movement of the strong cellulose microfibrils within the weaker hemicellulose/pectin matrix, catalyzed by expansin proteins. The outer part of the primary cell wall of the plant epidermis is usually impregnated with cutin and wax, forming a permeability barrier known as the plant cuticle.

Secondary cell walls contain a wide range of additional compounds that modify their mechanical properties and permeability. The major polymers that make up wood (largely secondary cell walls) include:

- cellulose, 35-50%

- xylan, 20-35%, a type of hemicellulose

- lignin, 10-25%, a complex phenolic polymer that penetrates the spaces in the cell wall between cellulose, hemicellulose and pectin components, driving out water and strengthening the wall.

Additionally, structural proteins (1-5%) are found in most plant cell walls; they are classified as hydroxyproline-rich glycoproteins (HRGP), arabinogalactan proteins (AGP), glycine-rich proteins (GRPs), and proline-rich proteins (PRPs). Each class of glycoprotein is defined by a characteristic, highly repetitive protein sequence. Most are glycosylated, contain hydroxyproline (Hyp) and become cross-linked in the cell wall. These proteins are often concentrated in specialized cells and in cell corners. Cell walls of the epidermis may contain cutin. The Casparian strip in the endodermis roots and cork cells of plant bark contain suberin. Both cutin and suberin are polyesters that function as permeability barriers to the movement of water. The relative composition of carbo-hydrates, secondary compounds and proteins varies between plants and between the cell type and age. Plant cells walls also contain numerous enzymes, such as hydrolases, esterases, peroxidases, and transglycosylases, that cut, trim and cross-link wall poly-mers.

Secondary walls - especially in grasses - may also contain microscopic silica crystals, which may strengthen the wall and protect it from herbivores.

Cell walls in some plant tissues also function as storage deposits for carbohydrates that can be broken down and resorbed to supply the metabolic and growth needs of the plant. For example, endosperm cell walls in the seeds of cereal grasses, nasturtium and other species, are rich in glucans and other polysaccharides that are readily digested by enzymes during seed germination to form simple sugars that nourish the growing embryo.

Formation

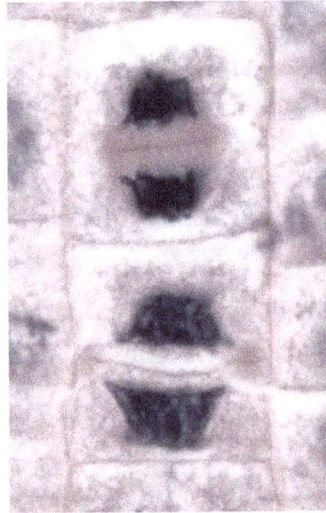

Photomicrograph of onion root cells, showing the centrifugal development of new cell walls (phragmoplast).

The middle lamella is laid down first, formed from the cell plate during cytokinesis, and the primary cell wall is then deposited inside the middle lamella. The actual structure of the cell wall is not clearly defined and several models exist - the covalently linked cross model, the tether model, the diffuse layer model and the stratified layer model. Howev-er, the primary cell wall, can be defined as composed of cellulosemicrofibrils aligned at all angles. Cellulose microfibrils are produced at the plasma membrane by the cellulose synthase complex, which is proposed to be made of a hexameric rosette that contains three cellulose synthase catalytic subunits for each of the six units. Microfibrils are held together by hydrogen bonds to provide a high tensile strength. The cells are held together and share the gelatinous membrane called the *middle lamella*, which contains magnesium and calciumpectates (salts of pectic acid). Cells interact though plasmodesmata, which are inter-connecting channels of cytoplasm that connect to the protoplasts of adjacent cells across the cell wall.

In some plants and cell types, after a maximum size or point in development has been reached, a *secondary wall* is constructed between the plasma membrane and primary wall. Unlike the primary wall, the cellulose microfibrils are aligned parallel in layers, the orientation changing slightly with each additional layer so that the structure becomes helicoidal. Cells with secondary cell walls can be rigid, as in the gritty sclereid cells in pear and quince fruit. Cell to cell communication is possible through pits in the secondary cell wall that allow plasmodesmata to connect cells through the secondary cell walls.

Fungal Cell Walls

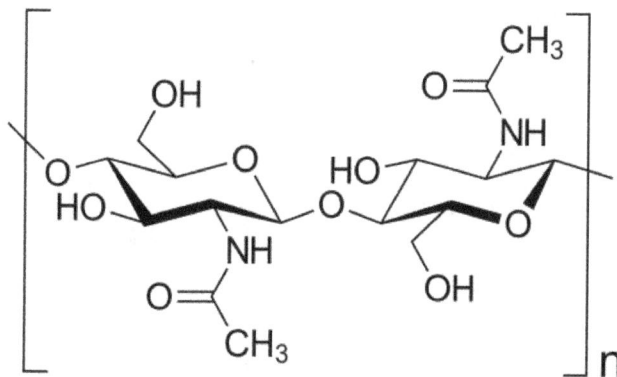

Chemical structure of a unit from a chitin polymer chain

There are several groups of organisms that have been called "fungi". Some of these groups (Oomycete and Myxogastria) have been transferred out of the Kingdom Fungi, in part because of fundamental biochemical differences in the composition of the cell wall. Most true fungi have a cell wall consisting largely of chitin and other polysaccharides. True fungi do not have cellulose in their cell walls.

True Fungi

In fungi, the cell wall is the outer-most layer, external to the plasma membrane. The fungal cell wall is a matrix of three main components:

- chitin: polymers consisting mainly of unbranched chains of β-(1,4)-l inked-N-Acetylglucosamine in the Ascomycota and Basidiomycota, or poly-β-(1,4)-linked-N-Acetylglucosamine (chitosan) in the Zygomycota. Both chitin and chitosan are synthesized and extruded at the plasma membrane.

- glucans: glucose polymers that function to cross-link chitin or chitosan polymers. β-glucans are glucose molecules linked via β-(1,3)- or β-(1,6)- bonds and provide rigidity to the cell wall while α-glucans are defined by α-(1,3)- and/or α-(1,4) bonds and function as part of the matrix.

- proteins: enzymes necessary for cell wall synthesis and lysis in addition to structural proteins are all present in the cell wall. Most of the structural proteins found in the cell wall are glycosylated and contain mannose, thus these proteins are called mannoproteins or mannans.

Other Eukaryotic Cell Walls

Algae

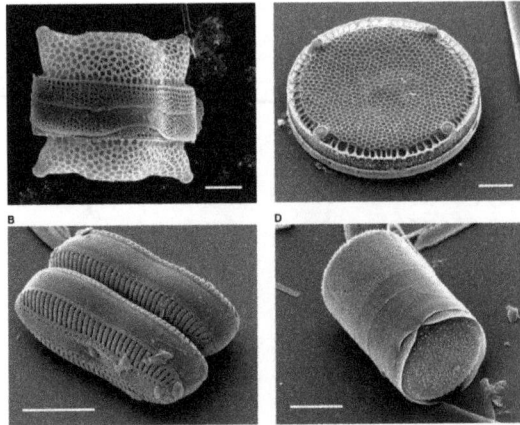

Scanning electronmicrographs of diatoms showing the external appearance of the cell wall

Like plants, algae have cell walls. Algal cell walls contain either polysaccharides (such as cellulose (a glucan)) or a variety of glycoproteins (Volvocales) or both. The inclusion of additional polysaccharides in algal cells walls is used as a feature for algal taxonomy.

- Mannans: They form microfibrils in the cell walls of a number of marine green algae including those from the genera, *Codium, Dasycladus*, and *Acetabularia* as well as in the walls of some red algae, like *Porphyra* and *Bangia*.

- Xylans.

- Alginic acid: It is a common polysaccharide in the cell walls of brown algae.

- Sulfonated polysaccharides: They occur in the cell walls of most algae; those common in red algae include agarose, carrageenan, porphyran, furcelleran and funoran.

Other compounds that may accumulate in algal cell walls include sporopollenin and calciumions.

The group of algae known as the diatoms synthesize their cell walls (also known as frustules or valves) from silicic acid (specifically orthosilicic acid, H_4SiO_4). The acid is polymerised intra-cellularly, then the wall is extruded to protect the cell. Significantly, relative to the organic cell walls produced by other groups, silica frustules require less energy to synthesize (approximately 8%), potentially a major saving on the overall cell energy budget and possibly an explanation for higher growth rates in diatoms.

In brown algae, phlorotannins may be a constituent of the cell walls.

Water Molds

The group Oomycetes, also known as water molds, are saprotrophicplant pathogens like fungi. Until recently they were widely believed to be fungi, but structural and molecular evidence has led to their reclassification as heterokonts, related to autotrophicbrown algae and diatoms. Unlike fungi, oomycetes typically possess cell walls of cellulose and glucans rather than chitin, although some genera (such as *Achlya* and *Saprolegnia*) do have chitin in their walls. The fraction of cellulose in the walls is no more than 4 to 20%, far less than the fraction of glucans. Oomycete cell walls also contain the amino acidhydroxyproline, which is not found in fungal cell walls.

Slime Molds

The dictyostelids are another group formerly classified among the fungi. They are slime molds that feed as unicellular amoebae, but aggregate into a reproductive stalk and sporangium under certain conditions. Cells of the reproductive stalk, as well as the spores formed at the apex, possess a cellulose wall. The spore wall has been shown to possess three layers, the middle of which is composed primarily of cellulose, and the innermost is sensitive to cellulase and pronase.

Prokaryotic Cell Walls

Bacterial Cell Walls

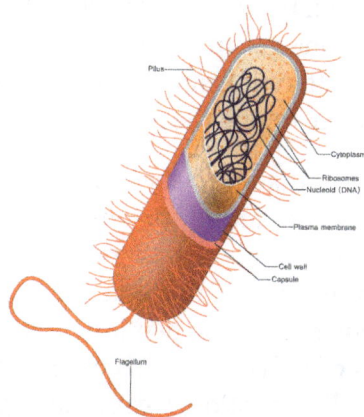

Diagram of a typical Gram-positive bacterium. The cell envelope comprises a plasma membrane, seen here in light brown, and a thick peptidoglycan-containing cell wall (the purple layer). No outer lipid membrane is present, as would be the case in Gram-negative bacteria. The red layer, known as the capsule, is distinct from the cell envelope.

Around the outside of the cell membrane is the bacterial cell wall. Bacterial cell walls are made of peptidoglycan (also called murein), which is made from polysaccharide chains cross-linked by unusual peptides containing D-amino acids. Bacterial cell walls are different from the cell walls of plants and fungi which are made of cellulose and chitin, respectively. The cell wall of bacteria is also distinct from that of Archaea, which do not contain peptidoglycan. The cell wall is essential to the survival of many bacteria, although L-form bacteria can be produced in the laboratory that lack a cell wall. The antibiotic penicillin is able to kill bacteria by preventing the cross-linking of peptidoglycan and this causes the cell wall to weaken and lyse. The lysozyme enzyme can also damage bacterial cell walls.

There are broadly speaking two different types of cell wall in bacteria, called Gram-positive and Gram-negative. The names originate from the reaction of cells to the Gram stain, a test long-employed for the classification of bacterial species.

Gram-positive bacteria possess a thick cell wall containing many layers of peptidoglycan and teichoic acids. In contrast, Gram-negative bacteria have a relatively thin cell wall consisting of a few layers of peptidoglycan surrounded by a second lipid membrane containing lipopolysaccharides and lipoproteins. Most bacteria have the Gram-negative cell wall and only the Firmicutes and Actinobacteria (previously known as the low G+C and high G+C Gram-positive bacteria, respectively) have the alternative Gram-positive arrangement. These differences in structure can produce differences in antibiotic susceptibility, for instance vancomycin can kill only Gram-positive bacteria and is ineffective against Gram-negative pathogens, such as *Haemophilus influenzae* or *Pseudomonas aeruginosa*.

Archaeal Cell Walls

Although not truly unique, the cell walls of Archaea are unusual. Whereas peptidoglycan is a standard component of all bacterial cell walls, all archaeal cell walls lack peptidoglycan, with the exception of one group of methanogens. In that group, the peptidoglycan is a modified form very different from the kind found in bacteria. There are four types of cell wall currently known among the Archaea.

One type of archaeal cell wall is that composed of pseudopeptidoglycan (also called pseudomurein). This type of wall is found in some methanogens, such as *Methanobacterium* and *Methanothermus*. While the overall structure of archaeal *pseudo*peptidoglycan superficially resembles that of bacterial peptidoglycan, there are a number of significant chemical differences. Like the peptidoglycan found in bacterial cell walls, pseudopeptidoglycan consists of polymer chains of glycan cross-linked by short peptide connections. However, unlike peptidoglycan, the sugar N-acetylmuramic acid is replaced by N-acetyltalosaminuronic acid, and the two sugars are bonded with a β,1-3 glycosidic linkage instead of β,1-4. Additionally, the cross-linking peptides are L-amino acids rather than D-amino acids as they are in bacteria.

A second type of archaeal cell wall is found in *Methanosarcina* and *Halococcus*. This type of cell wall is composed entirely of a thick layer of polysaccharides, which may be sulfated in the case of *Halococcus*. Structure in this type of wall is complex and not fully investigated.

A third type of wall among the Archaea consists of glycoprotein, and occurs in the hyperthermophiles, *Halobacterium*, and some methanogens. In *Halobacterium*, the proteins in the wall have a high content of acidicamino acids, giving the wall an overall negative charge. The result is an unstable structure that is stabilized by the presence of large quantities of positive sodiumions that neutralize the charge. Consequently, *Halobacterium* thrives only under conditions with high salinity.

In other Archaea, such as *Methanomicrobium* and *Desulfurococcus*, the wall may be composed only of surface-layer proteins, known as an *S-layer*. S-layers are common in bacteria, where they serve as either the sole cell-wall component or an outer layer in conjunction with polysaccharides. Most Archaea are Gram-negative, though at least one Gram-positive member is known.

Other Cell Coverings

Many protists and bacteria produce other cell surface structures apart from cell walls, external (extracellular matrix) or internal. Many algae have a sheath or envelope of mucilage outside the cell made of exopolysaccharides. Diatoms build a frustule from silica extracted from the surrounding water; radiolarians, foraminiferans, testate amoebae and silicoflagellates also produce a skeleton from minerals, called test in some groups. Many green algae, such as *Halimeda* and the Dasycladales, and some red algae, the Corallinales, encase their cells in a secreted skeleton of calcium carbonate. In each case, the wall is rigid and essentially inorganic. It is the non-living component of cell. Some golden algae, ciliates and choanoflagellatesproduces a shell-like protective outer covering called lorica. Some dinoflagellates have a theca of cellulose plates, and coccolithophorids have coccoliths.

An extracellular matrix is also present in metazoans. Its composition varies between cells, but collagens are the most abundant protein in the ECM.

Secondary Cell Wall

The secondary cell wall is a structure found in many plant cells, located between the primary cell wall and the plasma membrane. The cell starts producing the secondary cell wall after the primary cell wall is complete and the cell has stopped expanding.

Secondary cell walls provide additional protection to cells and rigidity and strength to the larger plant. These walls are constructed of layered sheaths of cellulose microfibrils, wherein the fibers are in parallel within each layer. The inclusion of lignin makes the secondary cell wall less flexible and less permeable to water than the primary cell wall. In addition to making the walls more resistant to degradation, the hydrophobic nature of lignin within these tissues is essential for containing water within the vascular tissues that carry it throughout the plant.

The secondary cell wall consists primarily of cellulose, along with other polysaccharides, lignin, and glycoprotein. It sometimes consists of three distinct layers - S_1, S_2 and S_3 - where the direction of the cellulose microfibrils differs between the layers.

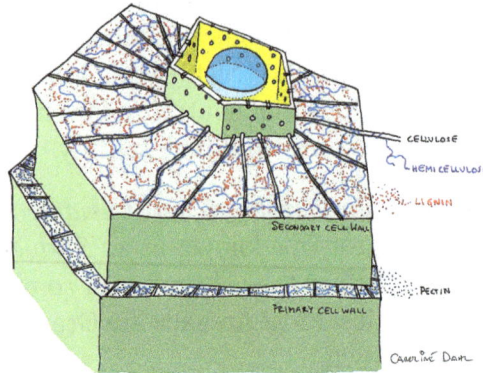

Plant cell overview, showing secondary cell wall.

The secondary cell wall has different ratios of constituents compared to the primary wall. An example of this is that secondary wall in wood contain polysaccharides called xylan, whereas the primary wall contains the polysaccharide xyloglucan. The cellulose fraction in secondary walls is also higher.Pectins may also be absent from the secondary wall, and unlike primary walls, no structural proteins or enzymes have been identified. Because of the low permeability through the secondary cell wall, cellular transport is carried out through openings in the wall called pits.

Wood consists mostly of secondary cell wall, and holds the plant up against gravity.

Some secondary cell walls store nutrients, such as those in the cotyledons and the endosperm. These contain little cellulose, and mostly other polysaccharides.

Acid Growth

Acid growth refers to the ability of plant cells and plant cell walls to elongate or expand quickly at low (acidic) pH. This form of growth does not involve an increase in cell number. During acid growth, plant cells enlarge rapidly because the cell walls are made more extensible by expansin, a pH-dependent wall-loosening protein. Expansin loosens the network-like connections between cellulose microfibrils within the cell wall, which allows the cell volume to increase by turgor and osmosis. A typical sequence leading up to this would involve the introduction of a plant hormone (auxin, for example) that causes protons (H^+ ions) to be pumped out of the cell into the cell wall. As a result, the cell wall solution becomes more acidic. This activates expansin activity, causing the wall to become more extensible and to undergo wall stress relaxation, which enables the cell to take up water and to expand.

Cellulose

Cellulose is an organic compound with the formula $(C_6H_{10}O_5)_n$, a polysaccharide consisting of a linear chain of several hundred to many thousands of β(1→4) linkedD-glucose units. Cellulose is an important structural component of the primary cell wall of green plants, many forms of algae and the oomycetes. Some species of bacteria secrete it to form biofilms. Cellulose is the most abundant organic polymer on Earth. The cellulose content of cotton fiber is 90%, that of wood is 40–50%, and that of dried hemp is approximately 57%.

Cellulose is mainly used to produce paperboard and paper. Smaller quantities are converted into a wide variety of derivative products such as cellophane and rayon. Conversion of cellulose from energy crops into biofuels such as cellulosic ethanol is under investigation as an alternative fuel source. Cellulose for industrial use is mainly obtained from wood pulp and cotton.

Some animals, particularly ruminants and termites, can digest cellulose with the help of symbiotic micro-organisms that live in their guts, such as *Trichonympha*. In human nutrition, cellulose is a non-digestible constituent of insolubledietary fiber, acting as a hydrophilicbulking agent for feces and potentially aiding in defecation.

History

Cellulose was discovered in 1838 by the French chemist Anselme Payen, who isolated it from plant matter and determined its chemical formula. Cellulose was used to produce the first successful thermoplastic polymer, celluloid, by Hyatt Manufacturing Company in 1870. Production of rayon ("artificial silk") from cellulose began in the 1890s and cellophane was invented in 1912. Hermann Staudinger determined the polymer structure of cellulose in 1920. The compound was first chemically synthesized (without the use of any biologically derived enzymes) in 1992, by Kobayashi and Shoda.

Structure and Properties

Cellulose has no taste, is odorless, is hydrophilic with the contact angle of 20–30 degrees, is insoluble in water and most organic solvents, is chiral and is biodegradable. It was shown to melt at 467 °C in 2016. It can be broken down chemically into its glucose units by treating it with concentrated mineral acids at high temperature.

Cellulose is derived from D-glucose units, which condense through β(1→4)-glycosidic bonds. This linkage motif contrasts with that for α(1→4)-glycosidic bonds present in starch and glycogen. Cellulose is a straight chain polymer: unlike starch, no coiling or branching occurs, and the molecule adopts an extended and rather stiff rod-like conformation, aided by the equatorial conformation of the glucose residues. The multiple hydroxyl groups on the glucose from one chain form hydrogen bonds with oxygen atoms on the same or on a neighbor chain, holding the chains firmly together side-by-side and forming *microfibrils* with high tensile strength. This confers tensile strength in cell walls, where cellulose microfibrils are meshed into a polysaccharide *matrix*.

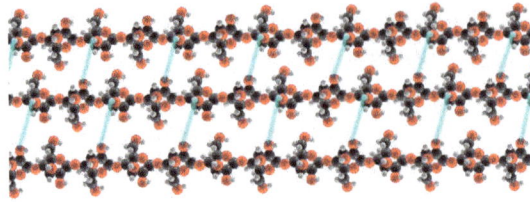

A triple strand of cellulose showing the hydrogen bonds (cyan lines) between glucose strands

Compared to starch, cellulose is also much more crystalline. Whereas starch undergoes a crystalline to amorphous transition when heated beyond 60–70 °C in water (as in cooking), cellulose requires a temperature of 320 °C and pressure of 25 MPa to become amorphous in water.

Cotton fibres represent the purest natural form of cellulose, containing more than 90% of this polysaccharide.

Several different crystalline structures of cellulose are known, corresponding to the location of hydrogen bonds between and within strands. Natural cellulose is cellulose I, with structures I_α and I_β. Cellulose produced by bacteria and algae is enriched in I_α while cellulose of higher plants consists mainly of I_β. Cellulose in regenerated cellulose fibers is cellulose II. The conversion of cellulose I to cellulose II is irreversible, suggesting that cellulose I ismetastable and cellulose II is stable. With various chemical treatments it is possible to produce the structures cellulose III and cellulose IV.

Many properties of cellulose depend on its chain length or degree of polymerization, the number of glucose units that make up one polymer molecule. Cellulose from wood pulp has typical chain lengths between 300 and 1700 units; cotton and other plant fibers as well as bacterial cellulose have chain lengths ranging from 800 to 10,000 units. Molecules with very small chain length resulting from the breakdown of cellulose are known as cellodextrins; in contrast to long-chain cellulose, cellodextrins are typically soluble in water and organic solvents.

Plant-derived cellulose is usually found in a mixture with hemicellulose, lignin, pectin and other substances, while bacterial cellulose is quite pure, has a much higher water content and higher tensile strength due to higher chain lengths.

Cellulose is soluble in Schweizer's reagent, cupriethylenediamine (CED), cadmiumethylenediamine (Cadoxen), N-methylmorpholine N-oxide, and lithium chloride / dimethylacetamide. This is used in the production of regenerated celluloses (such as viscose and cellophane) from dissolving pulp. Cellulose is also soluble in many kinds of ionic liquids.

Cellulose consists of crystalline and amorphous regions. By treating it with strong acid, the amorphous regions can be broken up, thereby producing nanocrystalline cellulose, a novel material with many desirable properties. Recently, nanocrystalline cellulose was used as the filler phase in bio-based polymer matrices to produce nanocomposites with superior thermal and mechanical properties.

Processing

Assay

Given a cellulose-containing material, the carbohydrate portion that does not dissolve in a 17.5% solution of sodium hydroxide at 20 °C is *α cellulose*, which is true cellulose. Acidification of the extract precipitates *β cellulose*. The portion that dissolves in base but does not precipitate with acid is *γ cellulose*.

Cellulose can be assayed using a method described by Updegraff in 1969, where the fiber is dissolved in acetic and nitric acid to remove lignin, hemicellulose, and xylosans. The resulting cellulose is allowed to react with anthrone in sulfuric acid. The resulting coloured compound is assayed spectrophotometrically at a wavelength of approximately 635 nm.

In addition, cellulose is represented by the difference between acid detergent fiber (ADF) and acid detergent lignin (ADL).

Luminescent conjugated oligothiophenes can also be used to detect cellulose using fluorescence microscopy or spectrofluorometric methods.

Biosynthesis

In vascular plants cellulose is synthesized at the plasma membrane by rosette terminal complexes (RTCs). The RTCs are hexameric protein structures, approximately 25 nm in diameter, that contain the cellulose synthase enzymes that synthesise the individual cellulose chains. Each RTC floats in the cell's plasma membrane and "spins" a microfibril into the cell wall.

RTCs contain at least three different cellulose synthases, encoded by *CesA* genes, in an unknown stoichiometry. Separate sets of *CesA* genes are involved in primary and secondary cell wall biosynthesis. There are known to be about seven subfamilies in the *CesA* superfamily. These cellulose synthases use UDP-glucose to form the β(1→4)-linked cellulose.

Cellulose synthesis requires chain initiation and elongation, and the two processes are separate. *CesA* glucosyltransferase initiates cellulose polymerization using a steroid primer, sitosterol-beta-glucoside, and UDP-glucose.Cellulose synthase utilizes UDP-D-glucose precursors to elongate the growing cellulose chain. A cellulase may function to cleave the primer from the mature chain.

Cellulose is also synthesised by animals, particularly in the tests of ascidians (where the cellulose was historically termed "tunicine") although it is also a minor component of mammalianconnective tissue.

Breakdown (Cellulolysis)

Cellulolysis is the process of breaking down cellulose into smaller polysaccharides called cellodextrins or completely into glucose units; this is a hydrolysis reaction. Because cellulose molecules bind strongly to each other, cellulolysis is relatively difficult compared to the breakdown of other polysaccharides. However, this process can be significantly intensified in a proper solvent, e.g. in an ionic liquid.

Most mammals have limited ability to digest dietary fiber such as cellulose. Some ruminants like cows and sheep contain certain symbioticanaerobic bacteria (like *Cellulomonas*) in the flora of the rumen, and these bacteria produce enzymes called cellulases that help the microorganism to digest cellulose; the breakdown products are then used by the bacteria for proliferation. The bacterial mass is later digested by the ruminant in its digestive system (stomach and small intestine). Horses use cellulose in their diet by fermentation in their hindgut via symbiotic bacteria which produce cellulase to digest cellulose. Similarly, some termites contain in their hindguts certain flagellateprotozoa producing such enzymes, whereas others contain bacteria or may produce cellulase.

The enzymes used to cleave the glycosidic linkage in cellulose are glycoside hydrolases including endo-acting cellulases and exo-acting glucosidases. Such enzymes are usually secreted as part of multienzyme complexes that may include dockerins and carbohydrate-binding modules.

Breakdown (Thermolysis)

At temperatures above 350 °C, cellulose undergoes thermolysis (also called 'pyrolysis'), decomposing into solid char, vapors, aerosols, and gases such as carbon dioxide. Maximum yield of vapors which condense to a liquid called *bio-oil* is obtained at 500 °C.

Semi-crystalline cellulose polymers react at pyrolysis temperatures (350–600 °C) in a few seconds; this transformation has been shown to occur via a solid-to-liquid-to-vapor transition, with the liquid (called *intermediate liquid cellulose* or *molten cellulose*) existing for only a fraction of a second. Glycosidic bond cleavage produces short cellulose chains of two-to-seven monomers comprising the melt. Vapor bubbling of intermediate liquid cellulose produces aerosols, which consist of short chain anhydro-oligomers derived from the melt.

Continuing decomposition of molten cellulose produces volatile compounds including levoglucosan, furans, pyrans, light oxygenates and gases via primary reactions. Within thick cellulose samples, volatile compounds such as levoglucosan undergo 'secondary reactions' to volatile products including pyrans and light oxygenates such as glycolaldehyde.

Hemicellulose

Hemicellulose is a polysaccharide related to cellulose that comprises about 20% of the biomass of most plants. In contrast to cellulose, hemicellulose is derived from several sugars in addition to glucose, especially xylose but also including mannose, galactose, rhamnose, and arabinose. Hemicellulose consists of shorter chains – between 500 and 3000 sugar units. Furthermore, hemicellulose is branched, whereas cellulose is unbranched.

Derivatives

The hydroxyl groups (-OH) of cellulose can be partially or fully reacted with various reagents to afford derivatives with useful properties like mainly cellulose esters and cellulose ethers (-OR). In principle, though not always in current industrial practice, cellulosic polymers are renewable resources.

Ester derivatives include:

Cellulose ester	Reagent	Example	Reagent	Group R
Organic esters	Organic acids	Cellulose acetate	Acetic acid and acetic anhydride	H or -(C=O)CH$_3$
		Cellulose triacetate	Acetic acid and acetic anhydride	-(C=O)CH$_3$
		Cellulose propionate	Propanoic acid	H or -(C=O)CH$_2$CH$_3$
		Cellulose acetate propionate (CAP)	Acetic acid and propanoic acid	H or -(C=O)CH$_3$ or -(C=O)CH$_2$CH$_3$
		Cellulose acetate butyrate (CAB)	Acetic acid and butyric acid	H or -(C=O)CH$_3$ or -(C=O)CH$_2$CH$_2$CH$_3$
Inorganic esters	Inorganic acids	Nitrocellulose (cellulose nitrate)	Nitric acid or another powerful nitrating agent	H or -NO$_2$
		Cellulose sulfate	Sulfuric acid or another powerful sulfuring agent	H or -SO$_3$H

The cellulose acetate and cellulose triacetate are film- and fiber-forming materials that find a variety of uses. The nitrocellulose was initially used as an explosive and was an early film forming material. With camphor, nitrocellulose gives celluloid.

Ether derivatives include:

Cellulose ethers	Reagent	Example	Reagent	Group R = H or	Water solubility	Application	E number
Alkyl	Halogenoalkanes	Methylcellulose	Chloromethane	$-CH_3$	Cold water-soluble		E461
		Ethylcellulose	Chloroethane	$-CH_2CH_3$	Water-insoluble	A commercial thermoplastic used in coatings, inks, binders, and controlled-release drug tablets	E462
		Ethyl methyl cellulose	Chloromethane and chloroethane	$-CH_3$ or $-CH_2CH_3$			E465
Hydroxyalkyl	Epoxides	Hydroxyethyl cellulose	Ethylene oxide	$-CH_2CH_2OH$	Cold/hot water-soluble	Gelling and thickening agent	
		Hydroxypropyl cellulose (HPC)	Propylene oxide	$-CH_2CH(OH)CH_3$	Cold water-soluble		E463
		Hydroxyethyl methyl cellulose	Chloromethane and ethylene oxide	$-CH_3$ or $-CH_2CH_2OH$	Cold water-soluble	Production of cellulose films	
		Hydroxypropyl methyl cellulose (HPMC)	Chloromethane and propylene oxide	$-CH_3$ or $-CH_2CH(OH)CH_3$	Cold water-soluble	Viscosity modifier, gelling, foaming and binding agent	E464
		Ethyl hydroxyethyl cellulose	Chloroethane and ethylene oxide	$-CH_2CH_3$ or $-CH_2CH_2OH$			E467
Carboxyalkyl	Halogenated carboxylic acids	Carboxymethyl cellulose (CMC)	Chloroacetic acid	$-CH_2COOH$	Cold/Hot water-soluble	Often used as its sodiumsalt, sodium carboxymethyl cellulose (NaCMC)	E466

The sodium carboxymethyl cellulose can be cross-linked to give the croscarmellose sodium (E468) for use as a disintegrant in pharmaceutical formulations.

Applications

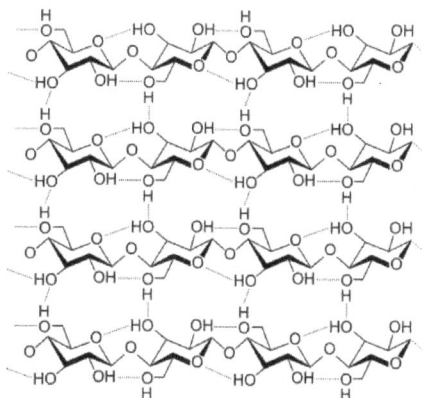

A strand of cellulose (conformation I$_\alpha$), showing the hydrogen bonds (dashed) within and between cellulose molecules.

Cellulose for industrial use is mainly obtained from wood pulp and cotton. The kraft process is used to separate cellulose from lignin, another major component of plant matter.

- Paper products: Cellulose is the major constituent of paper, paperboard, and card stock.

- Fibers: Cellulose is the main ingredient of textiles made from cotton, linen, and other plant fibers. It can be turned into rayon, an important fiber that has been used for textiles since the beginning of the 20th century. Both cellophane and rayon are known as "regenerated cellulose fibers"; they are identical to cellulose in chemical structure and are usually made from dissolving pulp via viscose. A more recent and environmentally friendly method to produce a form of rayon is the Lyocell process.

- Consumables: Microcrystalline cellulose (E460i) and powdered cellulose (E460ii) are used as inactive fillers in drug tablets and a wide range of soluble cellulose derivatives, E numbers E461 to E469, are used as emulsifiers, thickeners and stabilizers in processed foods. Cellulose powder is, for example, used in Parmesan cheese to prevent caking inside the package. Cellulose occurs naturally in some foods and is an additive in manufactured foods, contributing an indigestible component used for texture and bulk, potentially aiding in defecation.

- Science: Cellulose is used in the laboratory as a stationary phase for thin layer chromatography. Cellulose fibers are also used in liquid filtration, sometimes in combination with diatomaceous earth or other filtration media, to create a filter bed of inert material.

- Energy crops.

The major combustible component of non-food energy crops is cellulose, with lignin second. Non-food energy crops produce more usable energy than edible energy crops (which have a large starch component), but still compete with food crops for agricultural land and water resources. Typical non-food energy crops include industrial hemp (though outlawed in some countries), switchgrass, *Miscanthus*, *Salix* (willow), and *Populus* (poplar) species.

- Biofuel: TU-103, a strain of *Clostridium* bacteria found in zebra waste, can convert nearly any form of cellulose into butanol fuel.

- Building material: Hydroxyl bonding of cellulose in water produces a sprayable, moldable material as an alternative to the use of plastics and resins. The recyclable material can be made water- and fire-resistant. It provides sufficient strength for use as a building material.Cellulose insulation made from recycled paper is becoming popular as an environmentally preferable material for building insulation. It can be treated with boric acid as a fire retardant.

- Miscellaneous: Cellulose can be converted into cellophane, a thin transparent film. It is the base material for the celluloid that was used for photographic and movie films until the mid-1930s. Cellulose is used to make water-soluble adhesives and binders such as methyl cellulose and carboxymethyl cellulose which are used in wallpaper paste. Cellulose is further used to make hydrophilic and highly absorbent sponges. Cellulose is the raw material in the manufacture of nitrocellulose (cellulose nitrate) which is used in smokeless gunpowder.

Hemicellulose

A hemicellulose (also known as polyose) is any of several heteropolymers (matrix polysaccharides), such as arabinoxylans, present along with cellulose in almost all plant cell walls. While cellulose is crystalline, strong, and resistant to hydrolysis, hemicellulose has a random, amorphous structure with little strength. It is easily hydrolyzed by dilute acid or base as well as myriad hemicellulase enzymes.

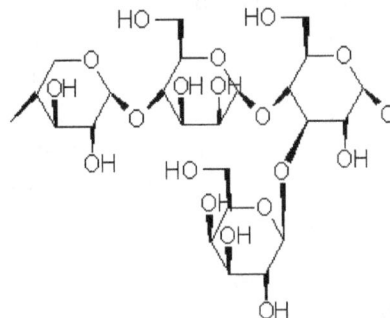

- Xylose - ß(1,4) - Mannose - ß(1,4) - Glucose -
- alpha(1,3) - Galactose

Hemicellulose
Most common molecular motif of hemicellulose

Composition

Hemicelluloses include xylan, glucuronoxylan, arabinoxylan, glucomannan, and xylo-glucan.

These polysaccharides contain many different sugar monomers. In contrast, cellulose contains only anhydrous glucose. For instance, besides glucose, sugar monomers in hemicellulose can include xylose, mannose, galactose, rhamnose, and arabinose. Hemicelluloses contain most of the D-pentose sugars, and occasionally small amounts of L-sugars as well. Xylose is in most cases the sugar monomer present in the largest amount, although in softwoods mannose can be the most abundant sugar. Not only regular sugars can be found in hemicellulose, but also their acidified form, for instance glucuronic acid and galacturonic acid can be present. Hemicellulose is often associated with cellulose, but it has different composition.

Structural Comparison to Cellulose

Unlike cellulose, hemicellulose (also a polysaccharide) consists of shorter chains – 500–3,000 sugar units as opposed to 7,000–15,000 glucose molecules per polymer, as seen in cellulose. In addition, hemicellulose is a branched polymer, while cellulose is unbranched.

Native Structure

Hemicelluloses are embedded in the cell walls of plants, sometimes in chains that form a 'ground' - they bind with pectin to cellulose to form a network of cross-linked fibres.

Biosynthesis

Hemicelluloses are synthesised from sugar nucleotides in the cell's Golgi apparatus. Two models explain their synthesis: 1) a '2 component model' where modification occurs at two transmembrane proteins, and 2) a '1 component model' where modification occurs only at one transmembrane protein. After synthesis, hemicelluloses are transported to the plasma membrane via Golgi vesicles.

Applications

As percent content of hemicellulose increases in animal feed, the voluntary feed intake decreases.

Hemicellulose is represented by the difference between neutral detergent fiber (NDF) and acid detergent fiber (ADF).

Functions

Microfibrils are cross-linked together by hemicellulose homopolymers. Lignins assist and strengthen the attachment of hemicelluloses to microfibrils.

Hemicellulose from Trees

Hemicellulose found in hardwood trees is predominantly xylan with some glucoman-nan, while in softwoods it is mainly rich in galactoglucomannan and contains only a small amount of xylan. The average molecular weight is lower than that of cellulose at less than 30,000, as opposed to the 100,000 average molecular weight reported for cellulose.

Pectin

Pectin, a bio polymer of (among other constituents)
D-galacturonic acid, shown here in a powder form.

Pectin is a structural heteropolysaccharide contained in the primary cell walls of ter-restrial plants. It was first isolated and described in 1825 by Henri Braconnot. It is produced commercially as a white to light brown powder, mainly extracted from citrus fruits, and is used in food as a gelling agent, particularly in jams and jellies. It is also used in dessert fillings, medicines, sweets, as a stabilizer in fruit juices and milk drinks, and as a source of dietary fiber.

Biology

In plant biology, pectin consists of a complex set of polysaccharides that are present in most primary cell walls and are particularly abundant in the non-woody parts of terres-trial plants. Pectin is a major component of the middle lamella, where it helps to bind cells together, but is also found in primary cell walls.

The amount, structure and chemical composition of pectin differs among plants, with-in a plant over time, and in various parts of a plant. Pectin is an important cell wall polysaccharide that allows primary cell wall extension and plant growth. During fruit ripening, pectin is broken down by the enzymespectinase and pectinesterase, in which process the fruit becomes softer as the middle lamellae break down and cells become separated from each other. A similar process of cell separation caused by the breakdown of pectin occurs in the abscission zone of the petioles of deciduous plants at leaf fall.

Pectin is a natural part of the human diet, but does not contribute significantly to nutrition. The daily intake of pectin from fruits and vegetables can be estimated to be around 5 g (assuming consumption of approximately 500 g fruits and vegetables per day).

In human digestion, pectin binds to cholesterol in the gastrointestinal tract and slows glucose absorption by trapping carbohydrates. Pectin is thus a soluble dietary fiber.

Pectin has been observed to have some function in DNA repair of plants. Pectinaceous surface pellicles, which are rich in pectin, create a mucilage layer that holds in dew that helps the cell repair its DNA.

Consumption of pectin has been shown to reduce blood cholesterol levels. The mechanism appears to be an increase of viscosity in the intestinal tract, leading to a reduced absorption of cholesterol from bile or food. In the large intestine and colon, microorganisms degrade pectin and liberate short-chain fatty acids that have positive influence on health (prebiotic effect).

Human Metabolites

A study found a mean of 4.5 ppmmethanol in the exhaled breath of subjects. The mean endogenous methanol production in humans of 0.45 g/d may be metabolized from pectin found in fruit; one kilogram of apple produces up to 1.4 g methanol.Methanol is poisonous to the central nervous system and may cause blindness, coma, and death. However, in small amounts, methanol is a natural endogenous compound found in normal, healthy human individuals.

Chemistry

Pectins, also known as pectic polysaccharides, are rich in galacturonic acid. Several distinct polysaccharides have been identified and characterised within the pectic group. Homogalacturonans are linear chains of α-(1–4)-linked D-galacturonic acid. Substituted galacturonans are characterized by the presence of saccharide appendant residues (such as D-xylose or D-apiose in the respective cases of xylogalacturonan and apiogalacturonan) branching from a backbone of D-galacturonic acid residues. Rhamnogalacturonan I pectins (RG-I) contain a backbone of the repeating disaccharide: 4)-α-D-galacturonic acid-(1,2)-α-L-rhamnose-(1. From many of the rhamnose residues, sidechains of various neutral sugars branch off. The neutral sugars are mainly D-galactose, L-arabinose and D-xylose, with the types and proportions of neutral sugars varying with the origin of pectin.

Another structural type of pectin is rhamnogalacturonan II (RG-II), which is a less frequent, complex, highly branched polysaccharide.Rhamnogalacturonan II is classified by some authors within the group of substituted galacturonans since the rhamnogalacturonan II backbone is made exclusively of D-galacturonic acid units.

Isolated pectin has a molecular weight of typically 60,000–130,000 g/mol, varying with origin and extraction conditions.

In nature, around 80 percent of carboxyl groups of galacturonic acid are esterified with methanol. This proportion is decreased to a varying degree during pectin extraction. The ratio of esterified to non-esterified galacturonic acid determines the behavior of pectin in food applications. This is why pectins are classified as high- vs. low-ester pectins (short HM vs. LM-pectins), with more or less than half of all the galacturonic acid esterified.

The non-esterified galacturonic acid units can be either free acids (carboxyl groups) or salts with sodium, potassium, or calcium. The salts of partially esterified pectins are called pectinates, if the degree of esterification is below 5 percent the salts are called pectates, the insoluble acid form, pectic acid.

Some plants such as sugar beet, potatoes and pears contain pectins with acetylated galacturonic acid in addition to methyl esters. Acetylation prevents gel-formation but increases the stabilising and emulsifying effects of pectin.

Amidated pectin is a modified form of pectin. Here, some of the galacturonic acid is converted with ammonia to carboxylic acidamide. These pectins are more tolerant of varying calcium concentrations that occur in use.

To prepare a pectin-gel, the ingredients are heated, dissolving the pectin. Upon cooling below gelling temperature, a gel starts to form. If gel formation is too strong, syneresis or a granular texture are the result, whilst weak gelling leads to excessively soft gels. Pectins gel according to specific parameters, such as sugar, pH and bivalent salts (especially Ca^{2+}).

In high-ester pectins at soluble solids content above 60% and a pH-value between 2.8 and 3.6, hydrogen bonds and hydrophobic interactions bind the individual pectin chains together. These bonds form as water is bound by sugar and forces pectin strands to stick together. These form a 3-dimensional molecular net that creates the macromolecular gel. The gelling-mechanism is called a low-water-activity gel or sugar-acid-pectin gel.

In low-ester pectins, ionic bridges are formed between calcium ions and the ionised carboxyl groups of the galacturonic acid. This is idealised in the so-called "egg box-model". Low-ester pectins need calcium to form a gel, but can do so at lower soluble solids and higher pH-values than high-ester pectins. Normally low-ester pectins form gels with a range of pH from 2.6 to 7.0 and with a soluble solids content between 10 and 70%.

Amidated pectins behave like low-ester pectins but need less calcium and are more tolerant of excess calcium. Also, gels from amidated pectin are thermo-reversible; they can be heated and after cooling solidify again, whereas conventional pectin-gels will afterwards remain liquid.

High-ester pectins set at higher temperatures than low-ester pectins. However, gelling reactions with calcium increase as the degree of esterification falls. Similarly, lower pH-values or higher soluble solids (normally sugars) increase gelling speed. Suitable pectins can therefore be selected for jams and for jellies, or for higher sugar confectionery jellies.

Sources and Production

Pears, apples, guavas, quince, plums, gooseberries, and oranges and other citrus fruits contain large amounts of pectin, while soft fruits like cherries, grapes, and strawberries contain small amounts of pectin.

Typical levels of pectin in plants are (fresh weight):

- apples, 1–1.5%
- apricots, 1%
- cherries, 0.4%
- oranges, 0.5–3.5%
- carrots approx. 1.4%
- citrus peels, 30%

The main raw materials for pectin production are dried citrus peel or apple pomace, both by-products of juice production. Pomace from sugar beet is also used to a small extent.

From these materials, pectin is extracted by adding hot dilute acid at pH-values from 1.5 – 3.5. During several hours of extraction, the protopectin loses some of its branching and chain length and goes into solution. After filtering, the extract is concentrated in vacuum and the pectin then precipitated by adding ethanol or isopropanol. An old technique of precipitating pectin with aluminium salts is no longer used (apart from alcohols and polyvalent cations, pectin also precipitates with proteins and detergents).

Alcohol-precipitated pectin is then separated, washed and dried. Treating the initial pectin with dilute acid leads to low-esterified pectins. When this process includes ammonium hydroxide, amidated pectins are obtained. After drying and milling, pectin is usually standardised with sugar and sometimes calcium salts or organic acids to have optimum performance in a particular application.

Uses

The main use for pectin (vegetable agglutinate) is as a gelling agent, thickening agent and stabilizer in food. The classical application is giving the jelly-like consistency to jams or marmalades, which would otherwise be sweet juices. Pectin also reduces syneresis in jams and marmalades and increases the gel strength of low calorie jams. For household use, pectin is an ingredient in gelling sugar (also known as "jam sugar") where it is diluted to the right concentration with sugar and some citric acid to adjust pH. In some countries, pectin is also available as a solution or an extract, or as a blended powder, for home jam making. For conventional jams and marmalades that contain above 60% sugar and soluble fruit solids, high-ester pectins are used. With low-ester pectins and amidated pectins less sugar is needed, so that diet products can be made.

Pectin is used in confectionery jellies to give a good gel structure, a clean bite and to confer a good flavour release. Pectin can also be used to stabilize acidic protein drinks, such as drinking yogurt, to improve the mouth-feel and the pulp stability in juice based drinks and as a fat substitute in baked goods. Typical levels of pectin used as a food additive are between 0.5 and 1.0% – this is about the same amount of pectin as in fresh fruit.

In medicine, pectin increases viscosity and volume of stool so that it is used against constipation and diarrhea. Until 2002, it was one of the main ingredients used in Kaopectate a medication to combat diarrhea, along with kaolinite. It has been used in gentle heavy metal removal from biological systems. Pectin is also used in throat lozenges as a demulcent.

In cosmetic products, pectin acts as stabilizer. Pectin is also used in wound healing preparations and specialty medical adhesives, such as colostomy devices.

Sriamornsak revealed that pectin could be used in various oral drug delivery platforms, e.g., controlled release systems, gastro-retentive systems, colon-specific delivery systems and mucoadhesive delivery systems, according to its intoxicity and low cost. It was found that pectin from different sources provides different gelling abilities, due to variations in molecular size and chemical composition. Like other natural polymers, a major problem with pectin is inconsistency in reproducibility between samples, which may result in poor reproducibility in drug delivery characteristics.

In ruminant nutrition, depending on the extent of lignification of the cell wall, pectin is up to 90% digestible by bacterial enzymes. Ruminant nutritionists recommend that the digestibility and energy concentration in forages can be improved by increasing pectin concentration in the forage.

In the cigar industry, pectin is considered an excellent substitute for vegetable glue and many cigar smokers and collectors will use pectin for repairing damaged tobacco wrapper leaves on their cigars.

Yablokov et al., writing in Chernobyl: Consequences of the Catastrophe for People and the Environment, quote research conducted by the Ukrainian Center of Radiation Medicine and the Belarusian Institute of Radiation Medicine and Endocrinology, concluded about pectin's radioprotective effects that "adding pectin preparations to the food of inhabitants of the Chernobyl-contaminated regions promotes an effective excretion of incorporated radionuclides" such as cesium-137. The authors report on the positive results of using pectin food additive preparations in a number of clinical studies conducted on children in severely polluted areas, with up to 50% improvement over control groups.

During the Second World War, Allied pilots were provided with maps printed on silk, for navigation in escape and evasion efforts. The printing process at first proved nearly impossible because the several layers of the ink immediately ran, blurring outlines and rendering place names illegible... until the inventor of the maps, Clayton Hutton, mixed a little pectin with the ink and at once the pectin coagulated the ink and prevented it from running. Now, even the smallest topographic features were sharply defined on the map.

Hazardous Metabolites

Fruit consumption is a source of endogenous methanol metabolized from pectin. One kilogram of apples produces up to 1.4 grams of methanol.

Legal Status

At the Joint FAO/WHO Expert Committee Report on Food Additives and in the European Union, no numerical acceptable daily intake (ADI) has been set, as pectin is considered safe.

In the United States, pectin is GRAS – generally recognized as safe. In most foods it can be used according to good manufacturing practices in the levels needed for its application ("quantum satis").

In the International Numbering System (INS), pectin has the number 440. In Europe, pectins are differentiated into the E numbersE440(i) for non-amidated pectins and E440(ii) for amidated pectins. There are specifications in all national and international legislation defining its quality and regulating its use.

History

Pectin was first isolated and described in 1825 by Henri Braconnot, though the action of pectin to make jams and marmalades was known long before. To obtain well set jams from fruits that had little or only poor quality pectin, pectin-rich fruits or their extracts were mixed into the recipe.

During the Industrial Revolution, the makers of fruit preserves turned to producers of apple juice to obtain dried apple pomace that was cooked to extract pectin.

Later, in the 1920s and 1930s, factories were built that commercially extracted pectin from dried apple pomace and later citrus-peel in regions that produced apple juice in both the USA and in Europe.

Pectin was first sold as a liquid extract, but is now most often used as dried powder, which is easier than a liquid to store and handle.

Middle Lamella

The middle lamella is a pectin layer which cements the cell walls of two adjoining plant cells together. It is the first formed layer which is deposited at the time of cytokinesis. The cell plate that is formed during cell division itself develops into middle lamella or lamellum. The middle lamella is made up of calcium and magnesium pectates. In a mature plant cell it is outermost layer of cell wall.

In plants, the pectins form a unified and continuous layer between adjacent cells. Frequently, it is difficult to distinguish the middle lamella from the primary wall, especially in cells that develop thick secondary walls. In such cases, the two adjacent primary walls and the middle lamella, and perhaps the first layer of the secondary wall of each cell, may be called a compound middle lamella. When the middle lamella is degraded by enzymes, as happens during fruit ripening, the adjacent cells will separate.

Lignin

Lignin is a class of complex organic polymers that form important structural materials in the support tissues of vascular plants and some algae. Lignins are particularly important in the formation of cell walls, especially in wood and bark, because they lend rigidity and do not rot easily. Chemically, lignins are cross-linked phenolic polymers.

History

Lignin was first mentioned in 1813 by the Swiss botanist A. P. de Candolle, who described it as a fibrous, tasteless material, insoluble in water and alcohol but soluble in weak alkaline solutions, and which can be precipitated from solution using acid. He named the substance "lignine", which is derived from the Latin word *lignum*, meaning wood. It is one of the most abundant organic polymers on Earth, exceeded only by cellulose. Lignin constitutes 30% of non-fossil organic carbon and 20-35% of the dry mass of wood. The CarboniferousPeriod (geology) is in part defined by the evolution of lignin.

Composition

The composition of lignin varies from species to species. An example of composition from an aspen sample is 63.4% carbon, 5.9% hydrogen, 0.7% ash (mineral components), and 30% oxygen (by difference), corresponding approximately to the formula $(C_{31}H_{34}O_{11})_n$. As a biopolymer, lignin is unusual because of its heterogeneity and lack of a defined primary structure. Its most commonly noted function is the support through strengthening of wood (mainly composed of xylem cells and lignified sclerenchyma fibres) in vascular plants.

Global commercial production of lignin is around 1.1 million metric tons per year and is used in a wide range of low volume, niche applications where the form but not the quality is important.

Biological Function

Lignin fills the spaces in the cell wall between cellulose, hemicellulose, and pectin components, especially in vascular and support tissues: xylemtracheids, vessel elements and sclereid cells. It is covalently linked to hemicellulose and therefore cross-links different plant polysaccharides, conferring mechanical strength to the cell wall and by extension the plant as a whole. It is particularly abundant in compression wood but scarce in tension wood, which are types of reaction wood.

Lignin plays a crucial part in conducting water in plant stems. The polysaccharide components of plant cell walls are highly hydrophilic and thus permeable to water, whereas lignin is more hydrophobic. The crosslinking of polysaccharides by lignin is an obstacle for water absorption to the cell wall. Thus, lignin makes it possible for the plant's vascular tissue to conduct water efficiently. Lignin is present in all vascular plants, but not in bryophytes, supporting the idea that the original function of lignin was restricted to water transport. However, it is present in red algae, which seems to suggest that the common ancestor of plants and red algae also synthesised lignin. This would suggest that its original function was structural; it plays this role in the red alga *Calliarthron*, where it supports joints between calcified segments. Another possibility is that the lignins in red algae and in plants are result of convergent evolution and not of a common origin.

Ecological Function

Lignin plays a significant role in the carbon cycle, sequestering atmospheric carbon into the living tissues of woody perennial vegetation. Lignin is one of the most slowly decomposing components of dead vegetation, contributing a major fraction of the material that becomes humus. The resulting soil humus, in general, holds nutrients onto its surface, and hence increases its cation exchange capacity and moisture retention, hence it increases the productivity of soil.

Economic Significance

Mechanical, or high-yield pulp used to make newsprint contains most of the lignin originally present in the wood. This lignin is responsible for newsprint's yellowing with age. Lignin must be removed from the pulp before high-quality bleached paper can be manufactured.

In sulfite pulping, lignin is removed from wood pulp as sulfonates. These lignosulfonates have several uses:

- Dispersants in high performance cement applications, water treatment formulations and textile dyes

- Additives in specialty oil field applications and agricultural chemicals

- Raw materials for several chemicals, such as vanillin, DMSO, ethanol, xylitol sugar, and humic acid

- Environmentally sustainable dust suppression agent for roads

Lignin removed via the kraft process is usually burned for its fuel value as part of a concentrated black liquor stream, providing energy to run the mill and its associated processes. Three commercial processes exist to remove lignin from black liquor for higher value uses: LignoBoost (Sweden), LignoForce (Canada), and SLRP (US). Higher quality lignin presents the potential to become the main renewable aromatic resource for the chemical industry in the future, with an addressable market of more than $130bn.

Given that lignin is the most prevalent biopolymer after cellulose and is ubiquitous in the Earth's biosphere, the same economic principles that drive the desire for cellulosic ethanol as a biofuel also call for the investigation of lignin as a feedstock for biofuel production. Lignin can already be burned in furnaces, but interest in the idea of instead chemically converting it to liquid fuel is strong.

Timeline

- 1927: The first investigations into commercial use of lignin were reported by Marathon Corporation, a paper company based in Rothschild, Wisconsin. The first class of products that showed promise were leather tanning agents. The lignin chemical business of Marathon was operated for many years as Marathon Chemicals. It is now known as LignoTech USA, Inc., and is owned by the Norwegian company Borregaard.

- 1998: a German company, Tecnaro, developed a process for turning lignin into a substance, called Arboform, which behaves identically to plastic for injection molding. Therefore, it can be used in place of plastic for several applications. When the item is discarded, it can be burned just like wood.

- 2007: lignin extracted from shrubby willow was successfully used to produce expanded polyurethane foam.

- 2012: it was shown carbon fiber can be produced from lignin instead of from fossil oil.

- 2013: the Flemish Institute for Biotechnology was supervising a trial of 448 poplar trees genetically engineered to produce less lignin so that they would be more suitable for conversion into bio-fuels.

- 2013: researchers from the University of California, Berkeley, demonstrated that lignin derived from a Miscanthus grass could be catalytically degraded into monophenolic products, which could potentially serve as an aromatic chemical feedstock.

Structure

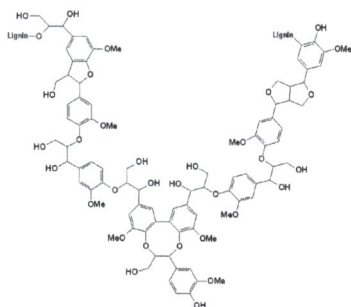

A small segment of lignin polymer

Lignin is a cross-linked racemicmacromolecule with molecular masses in excess of 10,000 u. It is relatively hydrophobic and aromatic in nature. The degree of polymerisation in nature is difficult to measure, since it is fragmented during extraction and the molecule consists of various types of substructures that appear to repeat in a haphazard manner. Different types of lignin have been described depending on the means of isolation.

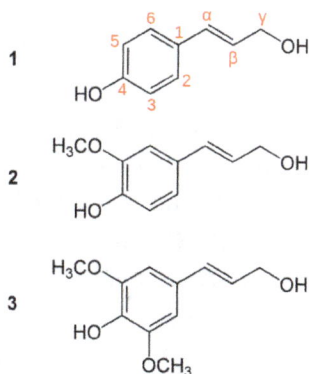

The three common monolignols: (1) paracoumaryl alcohol,
(2) coniferyl alcohol and (3) sinapyl alcohol.

There are three monolignolmonomers, methoxylated to various degrees: *p*-coumaryl alcohol, coniferyl alcohol, and sinapyl alcohol. These lignols are incorporated into lignin in the form of the phenylpropanoids*p*-hydroxyphenyl (H), guaiacyl (G), and syringyl (S), respectively.Gymnosperms have a lignin that consists almost entirely of G with small quantities of H. That of dicotyledonousangiosperms is more often than not a mixture of G and S (with very little H), and monocotyledonous lignin is a mixture of all three. Many grasses have mostly G, while some palms have mainly S. All lignins contain small amounts of incomplete or modified monolignols, and other monomers are prominent in non-woody plants.

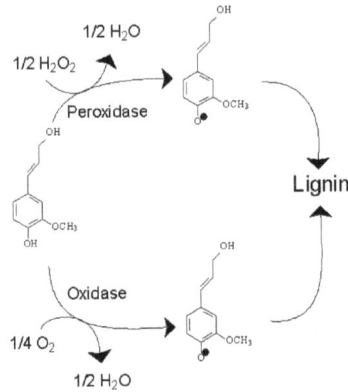

1/2 H_2O

1/2 H_2O_2

Peroxidase

Lignin

Oxidase

1/4 O_2

1/2 H_2O

Polymerisation of coniferyl alcohol to lignin. The reaction has two alternative routes catalysed by two different oxidative enzymes, peroxidases or oxidases.

Biosynthesis

Lignin biosynthesis begins in the cytosol with the synthesis of glycosylated monolignols from the amino acidphenylalanine. These first reactions are shared with the phenylpropanoid pathway. The attached glucose renders them water-soluble and less toxic. Once transported through the cell membrane to the apoplast, the glucose is removed and the polymerisation commences. Much about its anabolism is not understood even after more than a century of study.

The polymerisation step, that is a radical-radical coupling, is catalysed by oxidative enzymes. Both peroxidase and laccase enzymes are present in the plantcell walls, and it is not known whether one or both of these groups participates in the polymerisation. Low molecular weight oxidants might also be involved. The oxidative enzyme catalyses the formation of monolignol radicals. These radicals are often said to undergo uncatalyzed coupling to form the lignin polymer, but this hypothesis has been recently challenged. The alternative theory that involves an unspecified biological control is however not widely accepted.

Biodegradation

Biodegradation of lignin by white rot fungi leads to destruction of wood on the forest floor and human-made structures such as fences and wooden buildings. However biodegradation of lignin is a necessary prerequisite for processing biofuel from plant raw materials. Current processing setups show some problematic residuals after processing the digestible or degradable contents. The improving of lignin degradation would drive the output from biofuel processing to better gain or better efficiency.

Lignin is indigestible by animals, which lack the enzymes that can degrade this complex polymer. Some fungi (such as the Dryad's saddle) and bacteria do however biodegrade lignin using so-called ligninases (also named *lignases*). The mechanism of the biodegradation is speculated to involve free radical pathways. Well understood ligninolytic enzymes are manganese peroxidase and lignin peroxidase. Because it is cross-linked with the other cell wall components and has a high molecular weight, lignin minimizes the accessibility of cellulose and hemicellulose to microbial enzymes such as cellobiose dehydrogenase. Hence, in general lignin is associated with reduced digestibility of the overall plant biomass, which helps defend against pathogens and pests. Syringyl (S) lignol is more susceptible to degradation by fungal decay as it has fewer aryl-aryl bonds and a lower redox potential than guaiacyl units. This means that organic matter that is enriched with G lignol (like the bark of woody vascular plants) is more resistant to microbial attack.

Lignin is degraded by micro-organisms including fungi and bacteria. Lignin peroxidase (also "ligninase", EC number 1.14.99) is a hemoprotein firstly isolated from the white-rot fungus *Phanerochaete chrysosporium* with a variety of lignin-degrading reactions, all utilizing hydrogen peroxide as an oxygen source. Other microbial enzymes may be involved in lignin biodegradation, such as manganese peroxidase and the copper-based laccase.

Pyrolysis

Pyrolysis of lignin during the combustion of wood or charcoal production yields a range of products, of which the most characteristic ones are methoxy-substituted phenols. Of those, the most important are guaiacol and syringol and their derivatives; their presence can be used to trace a smoke source to a wood fire. In cooking, lignin in the form of hardwood is an important source of these two chemicals, which impart the characteristic aroma and taste to smoked foods such as barbecue. The main flavor compounds of smoked ham are guaiacol, and its 4-, 5-, and 6-methyl derivatives as well as 2,6-dimethylphenol. These compounds are produced by thermal breakdown of lignin in the wood used in the smokehouse.

Chemical Analysis

The conventional method for lignin quantitation in the pulp industry is the Klason lignin and acid-soluble lignin test, which is standardized according to SCAN or NREL procedure. The cellulose is first decrystallized and partially depolymerized into oligomers by keeping the sample in 72% sulfuric acid at 30 C for 1 h. Then, the acid is diluted to 4% by adding water, and the depolymerization is completed by either boiling (100 °C) for 4 h or pressure cooking at 2 bar (124 °C) for 1 h. The acid is washed out and the sample dried. The residue that remains is termed Klason lignin. A part of the lignin, acid-soluble lignin (ASL) dissolves in the acid. ASL is quantified by the intensity of its UV absorption peak at 280 nm. The method is suited for wood lignins, but not equally well for varied lignins from different sources. The carbohydrate composition may be also analyzed from the Klason liquors, although there may be sugar breakdown products (furfural and 5-hydroxymethylfurfural).

A solution of hydrochloric acid and phloroglucinol is used for the detection of lignin (Wiesner test). A brilliant red color develops, owing to the presence of coniferaldehyde groups in the lignin.

Thioglycolysis is an analytical technique for lignin quantitation. Lignin structure can also be studied by computational simulation.

Thermochemolysis (chemical break down of a substance under vacuum and at high temperature) with tetramethylammonium hydroxide (TMAH) has also been used to analyse the ratios of lignols with fungal decay as well the ratio of the carboxylic acid (Ad) to aldehyde (Al) forms of the lignols (Ad/Al). Increases in the (Ad/Al) value indicate an oxidative cleavage reaction has occurred on the alkyl lignin side chain which has been shown to be a step in the decay of wood by many white-rot and some soft rot fungi.

Solid state ^{13}C NMR has been used to look at the concentrations of lignin, as well as other major components in wood e.g. cellulose, and how that changes with microbial decay. Conventional solution-state NMR for lignin is possible. However, many intact lignins have a crosslinked, very high molar-mass fraction that is difficult to dissolve even for functionalization.

Plasmodesma

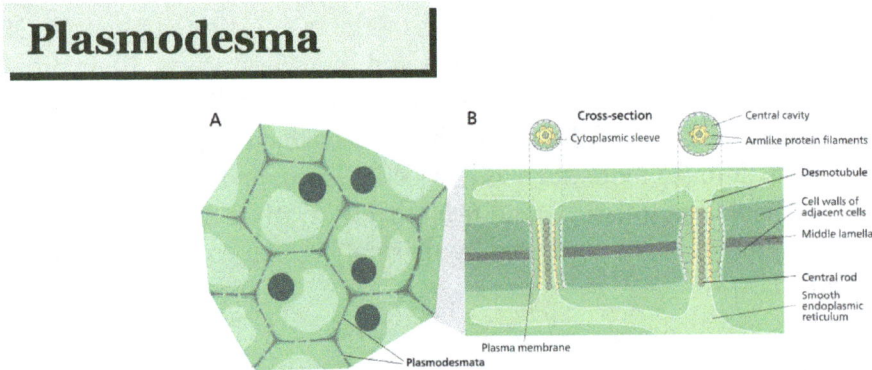

Diagram of some plasmodesmata

Plasmodesmata (singular: plasmodesma) are microscopic channels which traverse the cell walls of plant cells and some algal cells, enabling transport and communication between them. Plasmodesmata evolved independently in several lineages, and species that have these structures include members of the Charophyceae, Charales, Coleochaetales and Phaeophyceae (which are all algae), as well as all embryophytes, better known as land plants. Unlike animal cells, almost every plant cell is surrounded by a polysaccharidecell wall. Neighbouring plant cells are therefore separated by a pair of cell walls and the intervening middle lamella, forming an extracellular domain known as the apoplast. Although cell walls are permeable to small soluble proteins and other solutes, plasmodesmata enable direct, regulated, symplastic transport of substances between cells. There are two forms of plasmodesmata: primary plasmodesmata, which are formed during cell division, and secondary plasmodesmata, which can form between mature cells.

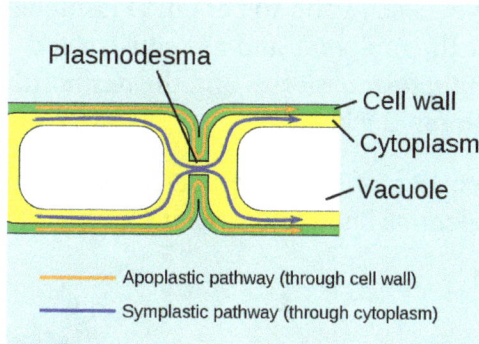

Plasmodesmata allow molecules to travel between plant cells through the symplastic pathway

Similar structures, called gap junctions and membrane nanotubes, interconnect animal cells and stromules form between plastids in plant cells.

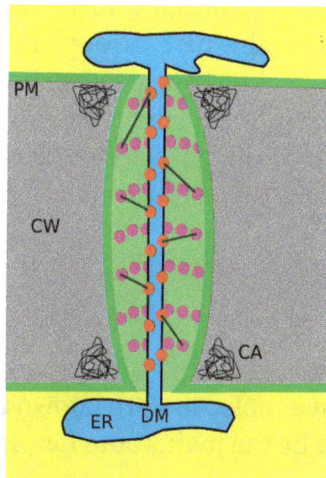

The structure of a primary plasmodesma. CW=Cell wall CA=Callose PM=Plasma membrane ER=Endoplasmic reticulum DM=Desmotubule Red circles=Actin Purple circles and spokes=Other unidentified proteins.

Formation

Primary plasmodesmata are formed when portions of the endoplasmic reticulum are trapped across the middle lamella as new cell wall is laid down between two newly divided plant cells and these eventually become the cytoplasmic connections between cells. Here the wall is not thickened further, and depressions or thin areas known as pits are formed in the walls. Pits normally pair up between adjacent cells. Plasmodesmata can also be inserted into existing cell walls between non-dividing cells (secondary plasmodesmata).

Structure

Plasmodesmatal Plasma Membrane

A typical plant cell may have between 10^3 and 10^5 plasmodesmata connecting it with adjacent cells equating to between 1 and 10 per μm^2. Plasmodesmata are approximately 50-60 nm in diameter at the midpoint and are constructed of three main layers, the plasma membrane, the *cytoplasmic sleeve*, and the *desmotubule*. They can transverse cell walls that are up to 90 nm thick.

The plasma membrane portion of the plasmodesma is a continuous extension of the cell membrane or plasmalemma and has a similar phospholipid bilayer structure.

Cytoplasmic Sleeve

The cytoplasmic sleeve is a fluid-filled space enclosed by the plasmalemma and is a continuous extension of the cytosol. Trafficking of molecules and ions through plasmodesmata occurs through this space. Smaller molecules (e.g. sugars and amino acids) and ions can easily pass through plasmodesmata by diffusion without the need for additional chemical energy. Larger molecules, including proteins (for example green fluorescent protein) and RNA, can also pass through the cytoplasmic sleeve diffusively. Plasmodesmatal transport of some larger molecules is facilitated by mechanisms that are currently unknown. One mechanism of regulation of the permeability of plasmodesmata is the accumulation of the polysaccharidecallose around the neck region to form a collar, thereby reducing the diameter of the pore available for transport of substances.

Desmotubule

The desmotubule is a tube of appressed (flattened) endoplasmic reticulum that runs between two adjacent cells. Some molecules are known to be transported through this channel, but it is not thought to be the main route for plasmodesmatal transport.

Around the desmotubule and the plasma membrane areas of an electron dense material have been seen, often joined together by spoke-like structures that seem to split the plasmodesma into smaller channels. These structures may be composed of myosin and actin, which are part of the cell's cytoskeleton. If this is the case these proteins could be used in the selective transport of large molecules between the two cells.

Transport

Plasmodesmata have been shown to transport proteins (including transcription factors), short interfering RNA, messenger RNA, viroids, and viral genomes from cell to cell. One example of a viral movement proteins is the tobacco mosaic virus MP-30. MP-30 is thought to bind to the virus's own genome and shuttle it from infected cells to uninfected cells through plasmodesmata.Flowering Locus T protein moves from leaves to the shoot apical meristem through plasmodesmata to initiate flowering.

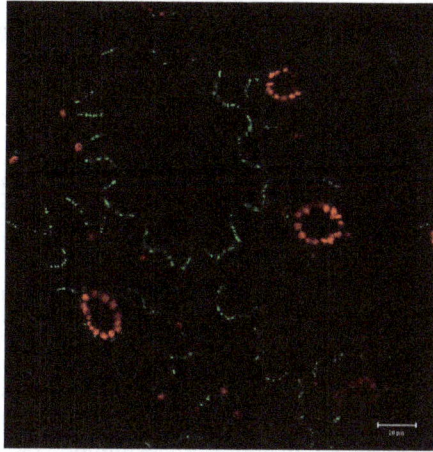

Tobacco mosaic virus movement protein 30 localizes to plasmodesmata.

Plasmodesmata are also used by cells in phloem, and symplastic transport is used to regulate the sieve-tube cells by the companion cells.

The size of molecules that can pass through plasmodesmata is determined by the size exclusion limit. This limit is highly variable and is subject to active modification. For example, MP-30 is able to increase the size exclusion limit from 700 Daltons to 9400 Daltons thereby aiding its movement through a plant. Also, increasing calcium concentrations in the cytoplasm, either by injection or by cold-induction, has been shown to constrict the opening of surrounding plasmodesmata and limit transport.

Several models for possible active transport through plasmodesmata exist. It has been suggested that such transport is mediated by interactions with proteins localized on the desmotubule, and/or by chaperones partially unfolding proteins, allowing them to fit through the narrow passage. A similar mechanism may be involved in transporting viral nucleic acids through the plasmodesmata.

References

- Endean, R (1961). "The Test of the Ascidian, Phallusia mammillata" (PDF). Quarterly Journal of Microscopical Science. 102 (1): 107–117

- Buchanan; Gruissem, Jones (2000). Biochemistry & molecular biology of plants (1st ed.). American society of plant physiology. ISBN 0-943088-39-9

- Sendbusch, Peter V. (2003-07-31). "Cell Walls of Algae Archived November 28, 2005, at the Wayback Machine.". Botany Online. Retrieved on 2007-10-29

- Raven, J. A. (1983). "The transport and function of silicon in plants". Biol. Rev. 58 (2): 179–207. doi:10.1111/j.1469-185X.1983.tb00385.x

- Howland, John L. (2000). The Surprising Archaea: Discovering Another Domain of Life. Oxford: Oxford University Press. pp. 69–71. ISBN 0-19-511183-4

- Brás, Natércia (2008). "Carbohydrate Binding Modules from family 11: Understanding the binding mode of polysaccharides". International Journal of Quantum Chemistry. 108 (11): 2030–2040. doi:10.1002/qua.21755

- Hobgood Ray, Kathryn (August 25, 2011). "Cars Could Run on Recycled Newspaper, Tulane Scientists Say". Tulane University news webpage. Tulane University. Retrieved March 14, 2012

- May, Colin D. (1990). "Industrial pectins: Sources, production and applications". Carbohydrate Polymers. 12 (1): 79–99. doi:10.1016/0144-8617(90)90105-2

- Dhingra, D; Michael, M; Rajput, H; Patil, R. T. (2011). "Dietary fibre in foods: A review". Journal of Food Science and Technology. 49 (3): 255–266. PMC 3614039. doi:10.1007/s13197-011-0365-5

- Stenius, Per (2000). "Ch. 1". Forest Products Chemistry. Papermaking Science and Technology. Vol. 3. Finland: Fapet OY. p. 35. ISBN 952-5216-03-9

- "Lignin and its Properties: Glossary of Lignin Nomenclature". Dialogue/Newsletters Volume 9, Number 1. Lignin Institute. July 2001. Retrieved 2007-10-14

- "Endogenous production of methanol after the consumption of fruit.". Alcohol Clin Exp Res. 21: 939–43. Aug 1997. PMID 9267548. doi:10.1111/j.1530-0277.1997.tb03862.x

- Gibson LJ (2013). "The hierarchical structure and mechanics of plant materials". Journal of the Royal Society Interface. 9 (76): 2749–2766. PMC 3479918. PMID 22874093. doi:10.1098/rsif.2012.0341

- Yablokov, Alexey V. Chernobyl Consequences of the Catastrophe for People and the Environment. John Wiley & Sons, 2010, pp. 304–309 ISBN 1573317578

- Richmond, Todd A; Somerville, Chris R (October 2000). "The Cellulose Synthase Superfamily". Plant Physiology. 124 (2): 495–498. PMC 1539280. PMID 11027699. doi:10.1104/pp.124.2.495. Retrieved 14 December 2014

- W. Boerjan; J. Ralph; M. Baucher (June 2003). "Lignin biosynthesis". Annu. Rev. Plant Biol. 54 (1): 519–549. PMID 14503002. doi:10.1146/annurev.arplant.54.031902.134938

- Sriamornsak, Pornsak (2003). "Chemistry of Pectin and its Pharmaceutical Uses: A Review". Silpakorn University International Journal. 3 (1–2): 206

- Lodish, Berk, Zipursky, Matsudaira, Baltimore, Darnell (2000). "22". Molecular Cell Biology (4 ed.). p. 998. ISBN 0-7167-3706-X. OCLC 41266312

- Blackman, LM; Overall, RL (1998). "Immunolocalisation of the cytoskeleton to plasmodesmata of Chara corallina". Plant Journal. 14: 733–741. doi:10.1046/j.1365-313x.1998.00161.x

- Radford, JE; White, RG (1998). "Localization of a myosin-like protein to plasmodesmata". Plant Journal. 14: 743–750. doi:10.1046/j.1365-313x.1998.00162.x

4

Plastids: An Integrated Study

The plastid is a double membrane organelle which contain pigments used in the process of photosynthesis. Gerontoplast, chromoplast, leucoplast and proteinoplast are some topics discussed in the chapter. This chapter provides a plethora of interdisciplinary topics for better comprehension of plastids.

Plastid

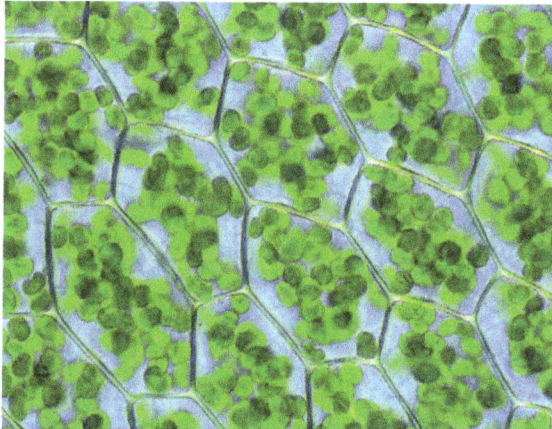
Plant cells with visible chloroplasts

The plastid is a major double-membrane organelle found, among others, in the cells of plants and algae. Plastids are the site of manufacture and storage of important chemical compounds used by the cell. They often contain pigments used in photosynthesis, and the types of pigments present can change or determine the cell's color. They have a common evolutionary origin and possess a double-stranded DNA molecule that is circular, like that of prokaryotic cells.

Plastids in Plants

Those plastids that contain chlorophyll can carry out photosynthesis. Plastids can also store products like starch and can synthesize fatty acids and terpenes, which can be used for producing energy and as raw material for the synthesis of other molecules. For example, the components of the plant cuticle and its epicuticular wax are synthesized by the epidermal cells from palmitic acid, which is synthesized in the chloroplasts of the mesophyll tissue. All plastids are derived from proplastids, which are present in the meristematic regions of the plant. Proplastids and young chloroplasts commonly divide by binary fission, but more mature chloroplasts also have this capacity.

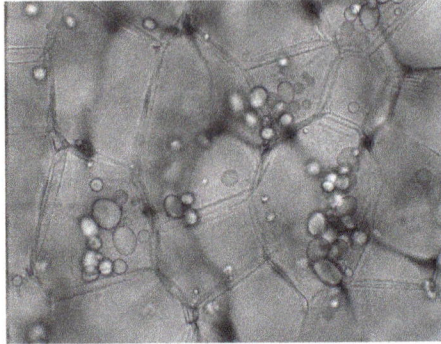

Leucoplasts in plant cells

In plants, plastids may differentiate into several forms, depending upon which function they play in the cell. Undifferentiated plastids (*proplastids*) may develop into any of the following variants:

- Chloroplasts green plastids: for photosynthesis; *the predecessors of chloro-plasts*

- Chromoplasts coloured plastids: for pigment synthesis and storage

- Gerontoplasts: control the dismantling of the photosynthetic apparatus during plant senescence

- Leucoplasts colourless plastids: for monoterpene synthesis; *leucoplasts sometimes differentiate into more specialized plastids:*

  o Amyloplasts: for starch storage and detecting gravity

  o Elaioplasts: for storing fat

  o Proteinoplasts: for storing and modifying protein

  o Tannosomes: for synthesizing and producing tannins and polyphenols

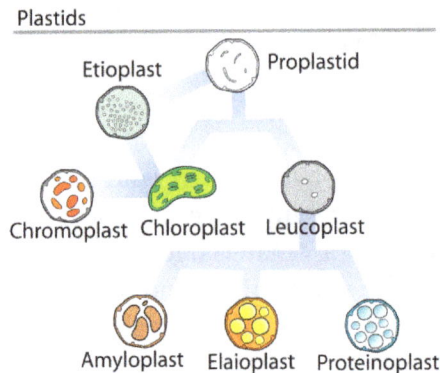

Depending on their morphology and function, plastids have the ability to differentiate, or redifferentiate, between these and other forms.

Each plastid creates multiple copies of a circular 75–250 kilobaseplastome. The number of genome copies per plastid is variable, ranging from more than 1000 in rapidly dividing cells, which, in general, contain few plastids, to 100 or fewer in mature cells, where plastid divisions have given rise to a large number of plastids. The plastome contains about 100 genes encoding ribosomal and transfer ribonucleic acids (rRNAs and tRNAs) as well as proteins involved in photosynthesis and plastid gene transcription and translation. However, these proteins only represent a small fraction of the total protein set-up necessary to build and maintain the structure and function of a particular type of plastid. Plant nuclear genes encode the vast majority of plastid proteins, and the expression of plastid genes and nuclear genes is tightly co-regulated to coordinate proper development of plastids in relation to cell differentiation.

Plastid DNA exists as large protein-DNA complexes associated with the inner envelope membrane and called 'plastid nucleoids'. Each nucleoid particle may contain more than 10 copies of the plastid DNA. The proplastid contains a single nucleoid located in the centre of the plastid. The developing plastid has many nucleoids, localized at the periphery of the plastid, bound to the inner envelope membrane. During the development of proplastids to chloroplasts, and when plastids convert from one type to another, nucleoids change in morphology, size and location within the organelle. The remodelling of nucleoids is believed to occur by modifications to the composition and abundance of nucleoid proteins.

Many plastids, particularly those responsible for photosynthesis, possess numerous internal membrane layers.

In plant cells, long thin protuberances called stromules sometimes form and extend from the main plastid body into the cytosol and interconnect several plastids. Proteins, and presumably smaller molecules, can move within stromules. Most cultured cells that are relatively large compared to other plant cells have very long and abundant stromules that extend to the cell periphery.

In 2014, evidence of possible plastid genome loss was found in *Rafflesia lagascae*, a non-photosynthetic parasitic flowering plant, and in *Polytomella*, a genus of non-photosynthetic green algae. Extensive searches for plastid genes in both *Rafflesia* and *Polytomella* yielded no results, however the conclusion that their plastomes are entirely missing is still controversial. Some scientists argue that plastid genome loss is unlikely since even non-photosynthetic plastids contain genes necessary to complete various biosynthetic pathways, such as heme biosynthesis.

Plastids in Algae

In algae, the term leucoplast is used for all unpigmented plastids and their function differs from the leucoplasts of plants. Etioplasts, amyloplasts and chromoplasts are plant-specific and do not occur in algae. Plastids in algae and hornworts may also differ from plant plastids in that they contain pyrenoids.

Glaucocystophytic algae contain muroplasts, which are similar to chloroplasts except that they have a peptidoglycan cell wall that is similar to that of prokaryotes. Rhodophytic algae contain rhodoplasts, which are red chloroplasts that allow the algae to photosynthesise to a depth of up to 268 m. The chloroplasts of plants differ from the rhodoplasts of red algae in their ability to synthesize starch, which is stored in the form of granules within the plastids. In red algae, floridean starch is synthesized and stored outside the plastids in the cytosol.

Inheritance of Plastids

Most plants inherit the plastids from only one parent. In general, angiosperms inherit plastids from the female gamete, whereas many gymnosperms inherit plastids from the male pollen. Algae also inherit plastids from only one parent. The plastid DNA of the other parent is, thus, completely lost.

In normal intraspecific crossings (resulting in normal hybrids of one species), the inheritance of plastid DNA appears to be quite strictly 100% uniparental. In interspecific hybridisations, however, the inheritance of plastids appears to be more erratic. Although plastids inherit mainly maternally in interspecific hybridisations, there are many reports of hybrids of flowering plants that contain plastids of the father. Approximately 20% of angiosperms, including alfalfa (*Medicago sativa*), normally show biparental inheritance of plastids.

Origin of Plastids

Plastids are thought to have originated from endosymbioticcyanobacteria. This symbiosis evolved around 1.5 billion years ago and enabled eukaryotes to carry out oxygenic photosynthesis. Three evolutionary lineages have since emerged in which the plastids are named differently: chloroplasts in green algae and plants, rhodoplasts in red algae and muroplasts in the glaucophytes. The plastids differ both in their pigmentation and in their ultrastructure. For example, chloroplasts have lost all phycobilisomes, the light harvesting complexes found in cyanobacteria, red algae and glaucophytes, but instead contain stroma and grana thylakoids, structures found only in plants and closely related green algae. The glaucocystophycean plastid — in contrast to chloroplasts and rhodoplasts — is still surrounded by the remains of the cyanobacterial cell wall. All these primary plastids are surrounded by two membranes.

Complex plastids start by secondary endosymbiosis (where a eukaryotic organism engulfs another eukaryotic organism that contains a primary plastid resulting in its endosymbiotic fixation), when a eukaryote engulfs a red or green alga and retains the algal plastid, which is typically surrounded by more than two membranes. In some cases these plastids may be reduced in their metabolic and/or photosynthetic capacity. Algae with complex plastids derived by secondary endosymbiosis of a red alga include the heterokonts, haptophytes, cryptomonads, and most dinoflagellates (= rhodoplasts). Those that endosymbiosed a green alga include the euglenids and chlorarachniophytes (= chloroplasts). The Apicomplexa, a phylum of obligate parasitic protozoa including the causative agents of malaria (*Plasmodium* spp.), toxoplasmosis (*Toxoplasma gondii*), and many other human or animal diseases also harbor a complex plastid (although this organelle has been lost in some apicomplexans, such as *Cryptosporidium parvum*, which causes cryptosporidiosis). The 'apicoplast' is no longer capable of photosynthesis, but is an essential organelle, and a promising target for antiparasitic drug development.

Some dinoflagellates and sea slugs, in particular of the genus *Elysia*, take up algae as food and keep the plastid of the digested alga to profit from the photosynthesis; after a while, the plastids are also digested. This process is known as kleptoplasty, from the Greek, *kleptes*, thief.

Chloroplast

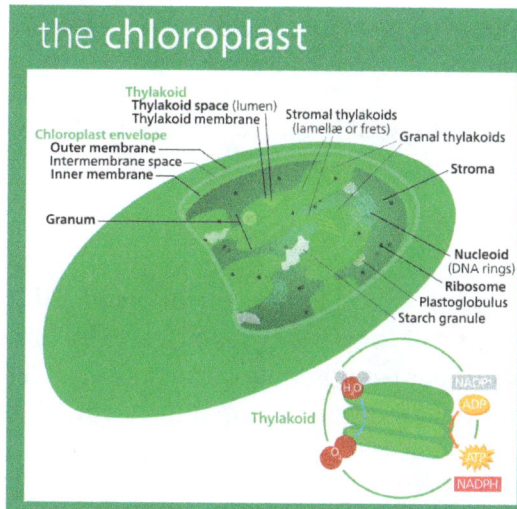

Structure of a typical higher-plant chloroplast

Chloroplasts are organelles, specialized subunits, in plant and algal cells. Their discovery inside plant cells is usually credited to Julius von Sachs (1832–1897), an influential botanist and author of standard botanical textbooks – sometimes called "The Father of Plant Physiology".

The main role of chloroplasts is to conduct photosynthesis, where the photosynthetic pigment chlorophyll captures the energy from sunlight and converts it and stores it in the energy-storage molecules ATP and NADPH while freeing oxygen from water. They then use the ATP and NADPH to make organic molecules from carbon dioxide in a process known as the Calvin cycle. Chloroplasts carry out a number of other functions, including fatty acid synthesis, much amino acid synthesis, and the immune response in plants. The number of chloroplasts per cell varies from one, in unicellular algae, up to 100 in plants like *Arabidopsis* and wheat.

A chloroplast is a type of organelle known as a plastid, characterized by its high concentration of chlorophyll. Other plastid types, such as the leucoplast and the chromoplast, contain little chlorophyll and do not carry out photosynthesis.

Chloroplasts are highly dynamic—they circulate and are moved around within plant cells, and occasionally pinch in two to reproduce. Their behavior is strongly influenced by environmental factors like light color and intensity. Chloroplasts, like mitochondria, contain their own DNA, which is thought to be inherited from their ancestor—a photosynthetic cyanobacterium that was engulfed by an early eukaryotic cell. Chloroplasts cannot be made by the plant cell and must be inherited by each daughter cell during cell division.

With one exception (the amoeboid*Paulinella chromatophora*), all chloroplasts can probably be traced back to a single endosymbiotic event, when a cyanobacterium was engulfed by the eukaryote. Despite this, chloroplasts can be found in an extremely wide set of organisms, some not even directly related to each other—a consequence of many secondary and even tertiary endosymbiotic events.

The word *chloroplast* is derived from the Greek words *chloros*, which means green, and *plastes*, which means the one who forms.

Discovery

The first definitive description of a chloroplast (*Chlorophyllkörnen*, "grain of chlorophyll") was given by Hugo von Mohl in 1837 as discrete bodies within the green plant cell. In 1883, A. F. W. Schimper would name these bodies as "chloroplastids" (*Chloroplastiden*). In 1884, Eduard Strasburger adopted the term "chloroplasts" (*Chloroplasten*).

Chloroplast Lineages and Evolution

Chloroplasts are one of many types of organelles in the plant cell. They are considered to have originated from cyanobacteria through endosymbiosis—when a eukaryotic cell engulfed a photosynthesizing cyanobacterium that became a permanent resident in the cell. Mitochondria are thought to have come from a similar event, where an aerobicprokaryote was engulfed. This origin of chloroplasts was first suggested by the Russian biologist Konstantin Mereschkowski in 1905 after Andreas Schimper observed in 1883 that chloroplasts closely resemble cyanobacteria. Chloroplasts are only found in plants, algae, and the amoeboid*Paulinella chromatophora.*

Cyanobacterial Ancestor

Cyanobacteria are considered the ancestors of chloroplasts. They are sometimes called blue-green algae even though they are prokaryotes. They are a diverse phylum of bacteria capable of carrying out photosynthesis, and are gram-negative, meaning that they have two cell membranes. Cyanobacteria also contain a peptidoglycan cell wall, which is thicker than in other gram-negative bacteria, and which is located between their two cell membranes. Like chloroplasts, they have thylakoids within. On the thylakoid membranes are photosynthetic pigments, including chlorophyll *a.*Phycobilins are also common cyanobacterial pigments, usually organized into hemispherical phycobilisomes attached to the outside of the thylakoid membranes (phycobilins are not shared with all chloroplasts though).

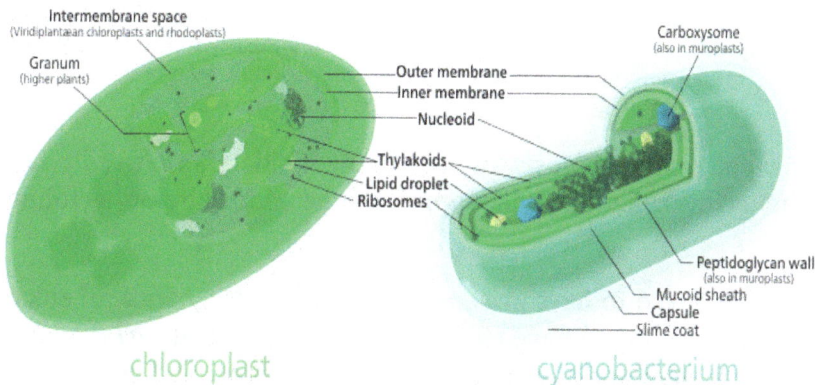

Both chloroplasts and cyanobacteria have a double membrane, DNA, ribosomes, and thylakoids. Both the chloroplast and cyanobacterium depicted are idealized versions (the chloroplast is that of a higher plant)—a lot of diversity exists among chloroplasts and cyanobacteria.

Primary Endosymbiosis

Somewhere around a billion years ago, a free-living cyanobacterium entered an early eukaryotic cell, either as food or as an internal parasite, but managed to escape the phagocytic vacuole it was contained in. The two innermost lipid-bilayer membranes that surround all chloroplasts correspond to the outer and inner membranes of the ancestral cyanobacterium's gram negative cell wall, and not the phagosomal membrane from the host, which was probably lost. The new cellular resident quickly became an advantage, providing food for the eukaryotic host, which allowed it to live within it. Over time, the cyanobacterium was assimilated, and many of its genes were lost or transferred to the nucleus of the host. From genomes that probably originally contained over 3000 genes only about 130 genes remain in the chloroplasts of contemporary plants. Some of its proteins were then synthesized in the cytoplasm of the host cell, and imported back into the chloroplast (formerly the cyanobacterium). Separately, somewhere around 100 million years ago, it happened again and led to the amoeboid-*Paulinella chromatophora*.

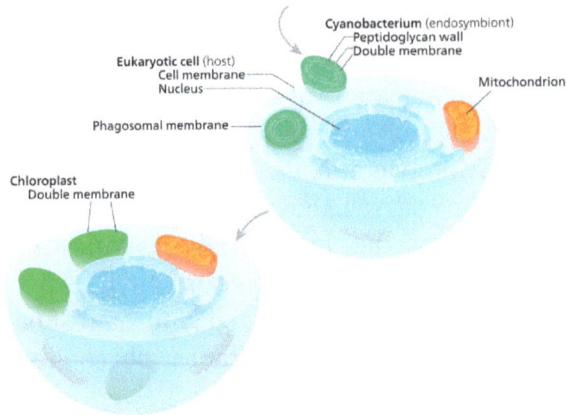

Primary endosymbiosis A eukaryote with mitochondria engulfed a cyanobacterium in an event of serial primary endosymbiosis, creating a lineage of cells with both organelles. It is important to note that the cyanobacterial endosymbiont already had a double membrane—
the phagosomal vacuole-derived membrane was lost.

This event is called *endosymbiosis*, or "cell living inside another cell". The cell living inside the other cell is called the *endosymbiont*; the endosymbiont is found inside the *host cell*.

Chloroplasts are believed to have arisen after mitochondria, since all eukaryotes contain mitochondria, but not all have chloroplasts. This is called *serial endosymbiosis*— an early eukaryote engulfing the mitochondrion ancestor, and some descendants of it then engulfing the chloroplast ancestor, creating a cell with both chloroplasts and mitochondria.

Whether or not chloroplasts came from a single endosymbiotic event, or many independent engulfments across various eukaryotic lineages, has been long debated. It is now generally held that most organisms with chloroplasts either share a single ancestor or obtained their chloroplast from organisms that share a common ancestor that took in a cyanobacterium 600–1600 million years ago. The exception is the amoeboid*Paulinella chromatophora*, which descends from an ancestor that took in a cyanobacterium 90–140 million years ago.

These chloroplasts, which can be traced back directly to a cyanobacterial ancestor, are known as *primary plastids* (*"plastid"* in this context means almost the same thing as chloroplast). All primary chloroplasts belong to one of four chloroplast lineages—the glaucophyte chloroplast lineage, the amoeboid*Paulinella chromatophora* lineage, the rhodophyte (red algal) chloroplast lineage, or the chloroplastidan (green) chloroplast lineage. The rhodophyte and chloroplastidan lineages are the largest, with chloroplastidan (green) being the one that contains the land plants.

Primary endosymbiosis	Secondary endosymbiosis	Tertiary endosymbiosis
Glaucophyta		**Chloroplast lineages**
Paulinella		A primary endosymbiosis event gave rise to four main lineages of chloroplasts in the glaucophytes, Paulinella, chlorophyta, and rhodophyta. Some of these algae were subsequently engulfed by other algae, becoming secondary (or tertiary) endosymbionts.
Chloroplastida	Euglenophyta	
Land plants	Chlorarachniophyta	
Green algae	Green algal dinophytes	
	Apicomplexa[a]	
Rhodophyceae (Red algae)	Peridinin-typedinophytes[b]	[a] The apicomplexans (malaria parasites), contain a red algal endosymbiont with a non-photosynthetic chloroplast. [b] 2–3 chloroplast membranes [a c] 2–4 chloroplast membranes
	Cryptophyta	
	Haptophyta	Haptophyte dinophytes[c]
	Heterokontophyta	Diatom dinophytes

Glaucophyta

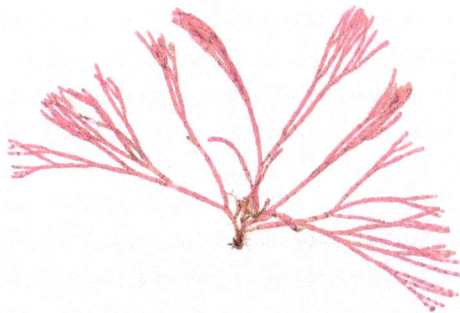

Bornetia secundiflora

The alga *Cyanophora*, a glaucophyte, is thought to be one of the first organisms to contain a chloroplast. The glaucophyte chloroplast group is the smallest of the three primary chloroplast lineages, being found in only 13 species, and is thought to be the one that branched off the earliest. Glaucophytes have chloroplasts that retain a peptidoglycan wall between their double membranes, like their cyanobacterial parent. For this reason, glaucophyte chloroplasts are also known as *muroplasts*. Glaucophyte chloroplasts also contain concentric unstacked thylakoids, which surround a carboxysome – an icosahedral structure that glaucophyte chloroplasts and cyanobacteria keep their carbon fixationenzymerubisco in. The starch that they synthesize collects outside the chloroplast. Like cyanobacteria, glaucophyte chloroplast thylakoids are studded with light collecting structures called phycobilisomes. For these reasons, glaucophyte chloroplasts are considered a primitive intermediate between cyanobacteria and the more evolved chloroplasts in red algae and plants.

Peyssonnelia squamaria

Cyanidium

Laurencia

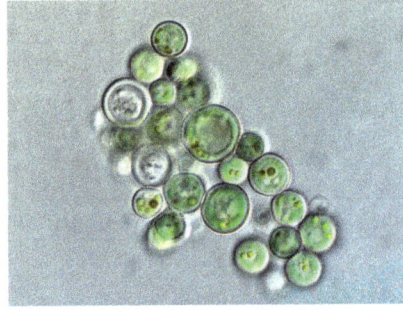
Callophyllis laciniata

Red algal chloroplasts are characterized by phycobilin pigments which often give them their reddish color.

Rhodophyceae (Red Algae)

The rhodophyte, or red algal chloroplast group is another large and diverse chloroplast lineage. Rhodophyte chloroplasts are also called *rhodoplasts*, literally "red chloroplasts".

Rhodoplasts have a double membrane with an intermembrane space and phycobilin pigments organized into phycobilisomes on the thylakoid membranes, preventing their thylakoids from stacking. Some contain pyrenoids. Rhodoplasts have chlorophyll *a* and phycobilins for photosynthetic pigments; the phycobilin phycoerytherin is responsible for giving many red algae their distinctive red color. However, since they also contain the blue-green chlorophyll *a* and other pigments, many are reddish to purple from the combination. The red phycoerytherin pigment is an adaptation to help red algae catch more sunlight in deep water—as such, some red algae that live in shallow water have less phycoerytherin in their rhodoplasts, and can appear more greenish. Rhodoplasts synthesize a form of starch called floridean starch, which collects into granules outside the rhodoplast, in the cytoplasm of the red alga.

Chloroplastida (Green Algae and Plants)

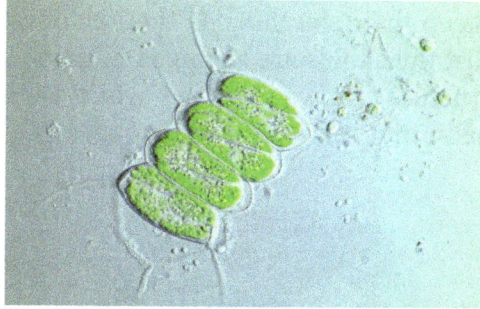

Scenedesmus

The chloroplastidan chloroplasts, or green chloroplasts, are another large, highly diverse primary chloroplast lineage. Their host organisms are commonly known as the green algae and land plants. They differ from glaucophyte and red algal chloroplasts in that they have lost their phycobilisomes, and contain chlorophyll *b* instead. Most green chloroplasts are (obviously) green, though some aren't, like some forms of *Hæmatococcus pluvialis*, due to accessory pigments that override the chlorophylls' green colors. Chloroplastidan chloroplasts have lost the peptidoglycan wall between their double membrane, leaving an intermembrane space. Some plants seem to have kept the genes for the synthesis of the peptidoglycan layer, though they've been repurposed for use in chloroplast division instead.

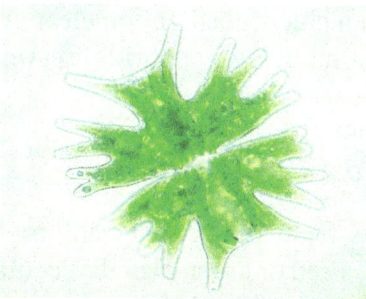

Micrasterias

Hydrodictyon

Volvox

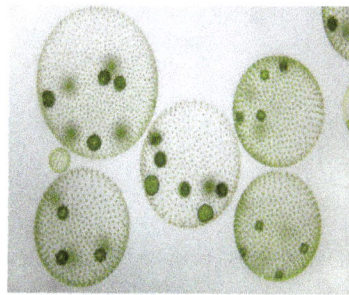

Stigeoclonium

Green algal chloroplasts are characterized by their pigments chlorophyll *a* and chlorophyll *b* which give them their green color.

Most of the chloroplasts depicted are green chloroplasts.

Green algae and plants keep their starch *inside* their chloroplasts, and in plants and some algae, the chloroplast thylakoids are arranged in grana stacks. Some green algal chloroplasts contain a structure called a pyrenoid, which is functionally similar to the glaucophyte carboxysome in that it is where rubisco and CO_2 are concentrated in the chloroplast.

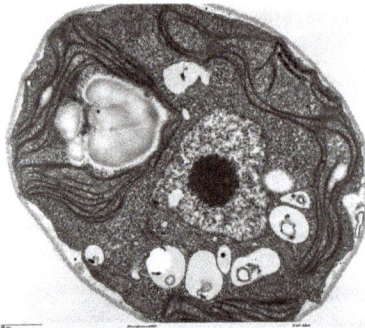

Transmission electron micrograph of *Chlamydomonas reinhardtii*, a green alga that contains a pyrenoid surrounded by starch.

Helicosporidium

Helicosporidium is a genus of nonphotosynthetic parasitic green algae that is thought to contain a vestigial chloroplast. Genes from a chloroplast and nuclear genes indicating the presence of a chloroplast have been found in Helicosporidium even if nobody's seen the chloroplast itself.

Secondary and Tertiary Endosymbiosis

Many other organisms obtained chloroplasts from the primary chloroplast lineages through secondary endosymbiosis—engulfing a red or green alga that contained a chloroplast. These chloroplasts are known as secondary plastids.

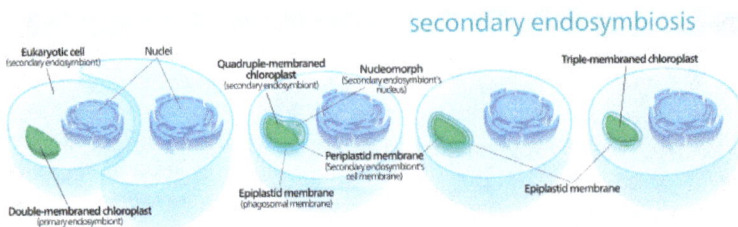

Secondary endosymbiosis consisted of a eukaryotic alga being engulfed by another eukaryote, forming a chloroplast with three or four membranes.

While primary chloroplasts have a double membrane from their cyanobacterial ancestor, secondary chloroplasts have additional membranes outside of the original two, as a result of the secondary endosymbiotic event, when a nonphotosynthetic eukaryote engulfed a chloroplast-containing alga but failed to digest it—much like the cyanobacterium at the beginning of this story. The engulfed alga was broken down, leaving only its chloroplast, and sometimes its cell membrane and nucleus, forming a chloroplast with three or four membranes—the two cyanobacterial membranes, sometimes the eaten alga's cell membrane, and the phagosomal vacuole from the host's cell membrane.

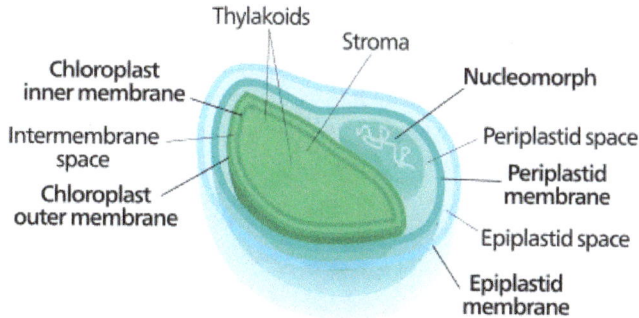

Diagram of a four membraned chloroplast containing a nucleomorph

The genes in the phagocytosed eukaryote's nucleus are often transferred to the secondary host's nucleus.Cryptomonads and chlorarachniophytes retain the phagocytosed eukaryote's nucleus, an object called a nucleomorph, located between the second and third membranes of the chloroplast.

All secondary chloroplasts come from green and red algae—no secondary chloroplasts from glaucophytes have been observed, probably because glaucophytes are relatively rare in nature, making them less likely to have been taken up by another eukaryote.

Green Algal Derived Chloroplasts

Green algae have been taken up by the euglenids, chlorarachniophytes, a lineage of dinoflagellates, and possibly the ancestor of the chromalveolates in three or four separate engulfments. Many green algal derived chloroplasts contain pyrenoids, but unlike chloroplasts in their green algal ancestors, starch collects in granules outside the chloroplast.

Euglena, a euglenophyte, contains secondary chloroplasts from green algae

Euglenophytes

Euglenophytes are a group of common flagellatedprotists that contain chloroplasts derived from a green alga.Euglenophyte chloroplasts have three membranes—it is thought that the membrane of the primary endosymbiont was lost, leaving the cyanobacterial membranes, and the secondary host's phagosomal membrane. Euglenophyte chloroplasts have a pyrenoid and thylakoids stacked in groups of three. Starch is stored in the form of paramylon, which is contained in membrane-bound granules in the cytoplasm of the euglenophyte.

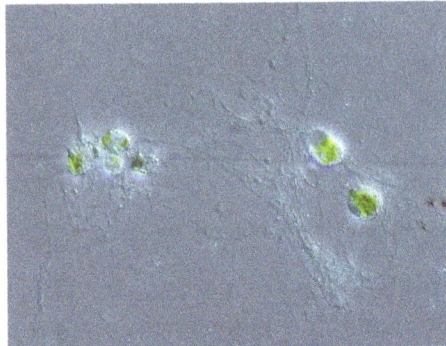

Chlorarachnion reptans is a chlorarachniophyte. Chlorarachniophytes replaced their original red algal endosymbiont with a green alga.

Chlorarachniophytes

Chlorarachniophytes are a rare group of organisms that also contain chloroplasts derived from green algae, though their story is more complicated than that of the euglenophytes. The ancestor of chlorarachniophytes is thought to have been a chromalveolate, a eukaryote with a *red* algal derived chloroplast. It is then thought to have lost its first red algal chloroplast, and later engulfed a green alga, giving it its second, green algal derived chloroplast.

Chlorarachniophyte chloroplasts are bounded by four membranes, except near the cell membrane, where the chloroplast membranes fuse into a double membrane. Their thylakoids are arranged in loose stacks of three. Chlorarachniophytes have a form of starch called chrysolaminarin, which they store in the cytoplasm, often collected around the chloroplast pyrenoid, which bulges into the cytoplasm.

Chlorarachniophyte chloroplasts are notable because the green alga they are derived from has not been completely broken down—its nucleus still persists as a nucleomorph found between the second and third chloroplast membranes—the periplastid space, which corresponds to the green alga's cytoplasm.

Early Chromalveolates

Recent research has suggested that the ancestor of the chromalveolates acquired a green algal prasinophyte endosymbiont. The green algal derived chloroplast was lost and replaced with a red algal derived chloroplast, but not before contributing some of its genes to the early chromalveolate's nucleus. The presence of both green algal and red algal genes in chromalveolates probably helps them thrive under fluctuating light conditions.

Red Algal Derived Chloroplasts (Chromalveolate Chloroplasts)

Like green algae, red algae have also been taken up in secondary endosymbiosis, though it is thought that all red algal derived chloroplasts are descended from a single red alga that was engulfed by an early chromalveolate, giving rise to the chromalveolates, some of which, like the ciliates, subsequently lost the chloroplast. This is still debated though.

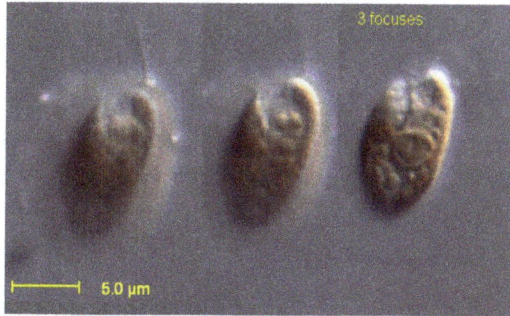

Rhodomonas salina is a cryptophyte

Pyrenoids and stacked thylakoids are common in chromalveolate chloroplasts, and the outermost membrane of many are continuous with the rough endoplasmic reticulum and studded with ribosomes. They have lost their phycobilisomes and exchanged them for chlorophyll *c*, which isn't found in primary red algal chloroplasts themselves.

Cryptophytes

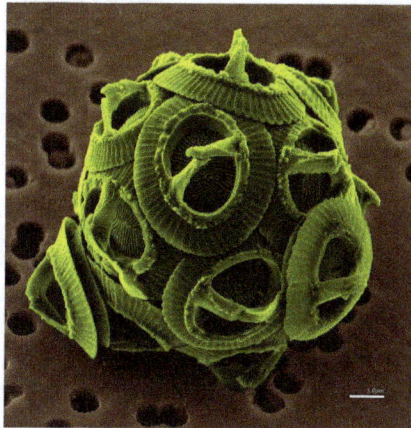

Scanning electron micrograph of *Gephyrocapsa oceanica*, a haptophyte.

Cryptophytes, or cryptomonads are a group of algae that contain a red-algal derived chloroplast. Cryptophyte chloroplasts contain a nucleomorph that superficially resembles that of the chlorarachniophytes. Cryptophyte chloroplasts have four membranes, the outermost of which is continuous with the rough endoplasmic reticulum. They synthesize ordinary starch, which is stored in granules found in the periplastid space—outside the original double membrane, in the place that corresponds to the red alga's cytoplasm. Inside cryptophyte chloroplasts is a pyrenoid and thylakoids in stacks of two.

Their chloroplasts do not have phycobilisomes, but they do have phycobilin pigments which they keep in their thylakoid space, rather than anchored on the outside of their thylakoid membranes.

Haptophytes

Haptophytes are similar and closely related to cryptophytes, and are thought to be the first chromalveolates to branch off. Their chloroplasts lack a nucleomorph, their thylakoids are in stacks of three, and they synthesize chrysolaminarin sugar, which they store completely outside of the chloroplast, in the cytoplasm of the haptophyte.

Heterokontophytes (Stramenopiles)

The photosynthetic pigments present in their chloroplasts give diatoms a greenish-brown color

The heterokontophytes, also known as the stramenopiles, are a very large and diverse group of algae that also contain red algal derived chloroplasts. Heterokonts include the diatoms and the brown algae, golden algae, and yellow-green algae.

Heterokont chloroplasts are very similar to haptophyte chloroplasts, containing a pyrenoid, triplet thylakoids, and with some exceptions, having an epiplastid membrane connected to the endoplasmic reticulum. Like haptophytes, heterokontophytes store sugar in chrysolaminarin granules in the cytoplasm. Heterokontophyte chloroplasts contain chlorophyll a and with a few exceptionschlorophyll c, but also have carotenoids which give them their many colors.

Apicomplexans

Apicomplexans are another group of chromalveolates. Like the helicosproidia, they're parasitic, and have a nonphotosynthetic chloroplast. They were once thought to be related to the helicosproidia, but it is now known that the helicosproida are green algae rather than chromalveolates. The apicomplexans include *Plasmodium*, the malaria parasite. Many apicomplexans keep a vestigial red algal derived chloroplast called an apicoplast, which they inherited from their ancestors. Other apicomplexans like *Cryptosporidium* have lost the chloroplast completely. Apicomplexans store their energy in amylopectin starch granules that are located in their cytoplasm, even though they are nonphotosynthetic.

Apicoplasts have lost all photosynthetic function, and contain no photosynthetic pigments or true thylakoids. They are bounded by four membranes, but the membranes are not connected to the endoplasmic reticulum. The fact that apicomplexans still keep their nonphotosynthetic chloroplast around demonstrates how the chloroplast carries out important functions other than photosynthesis. Plant chloroplasts provide plant cells with many important things besides sugar, and apicoplasts are no different—they synthesize fatty acids, isopentenyl pyrophosphate, iron-sulfur clusters, and carry out part of the heme pathway. This makes the apicoplast an attractive target for drugs to cure apicomplexan-related diseases.The most important apicoplast function is isopentenyl pyrophosphate synthesis—in fact, apicomplexans die when something interferes with this apicoplast function, and when apicomplexans are grown in an isopentenyl pyrophosphate-rich medium, they dump the organelle.

Dinophytes

The dinoflagellates are yet another very large and diverse group of protists, around half of which are (at least partially) photosynthetic.

Most dinophyte chloroplasts are secondary red algal derived chloroplasts, like other chromalveolate chloroplasts. Many other dinophytes have lost the chloroplast (becoming the nonphotosynthetic kind of dinoflagellate), or replaced it though *tertiary*endosymbiosis—the engulfment of another chromalveolate containing a red algal derived chloroplast. Others replaced their original chloroplast with a green algal derived one.

Most dinophyte chloroplasts contain at least the photosynthetic pigmentschlorophyll *a*, chlorophyll c_2, *beta*-carotene, and at least one dinophyte-unique xanthophyll (peridinin, dinoxanthin, or diadinoxanthin), giving many a golden-brown color. All dinophytes store starch in their cytoplasm, and most have chloroplasts with thylakoids arranged in stacks of three.

Peridinin-containing Dinophyte Chloroplast

The most common dinophyte chloroplast is the peridinin-type chloroplast, characterized by the carotenoid pigment peridinin in their chloroplasts, along with chlorophyll *a* and chlorophyll c_2. Peridinin is not found in any other group of chloroplasts. The peridinin chloroplast is bounded by three membranes (occasionally two), having lost the red algal endosymbiont's original cell membrane. The outermost membrane is not connected to the endoplasmic reticulum. They contain a pyrenoid, and have triplet-stacked thylakoids. Starch is found outside the chloroplast. An important feature of these chloroplasts is that their chloroplast DNA is highly reduced and fragmented into many small circles. Most of the genome has migrated to the nucleus, and only critical photosynthesis-related genes remain in the chloroplast.

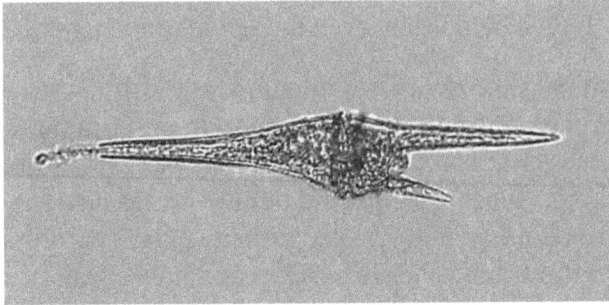

Ceratium furca, a peridinin-containing dinophyte

The peridinin chloroplast is thought to be the dinophytes' "original" chloroplast, which has been lost, reduced, replaced, or has company in several other dinophyte lineages.

Fucoxanthin-containing Dinophyte Chloroplasts (Haptophyte Endosymbionts)

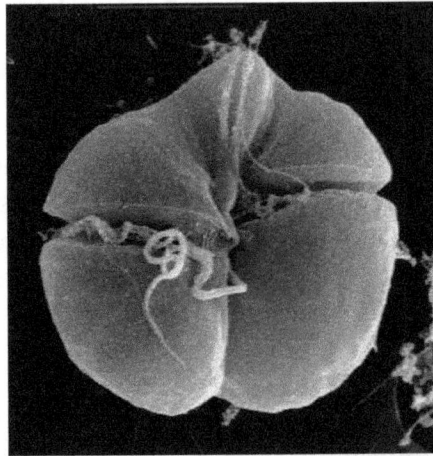

Karenia brevis is a fucoxanthin-containing dynophyte responsible for algal blooms called "red tides"

The fucoxanthin dinophyte lineages (including *Karlodinium* and *Karenia*) lost their original red algal derived chloroplast, and replaced it with a new chloroplast derived from a haptophyte endosymbiont. *Karlodinium* and *Karenia* probably took up different heterokontophytes. Because the haptophyte chloroplast has four membranes, tertiary endosymbiosis would be expected to create a six membraned chloroplast, adding the haptophyte's cell membrane and the dinophyte's phagosomal vacuole. However, the haptophyte was heavily reduced, stripped of a few membranes and its nucleus, leaving only its chloroplast (with its original double membrane), and possibly one or two additional membranes around it.

Fucoxanthin-containing chloroplasts are characterized by having the pigment fucoxanthin (actually 19'-hexanoyloxy-fucoxanthin and/or 19'-butanoyloxy-fucoxanthin) and no peridinin. Fucoxanthin is also found in haptophyte chloroplasts, providing evidence of ancestry.

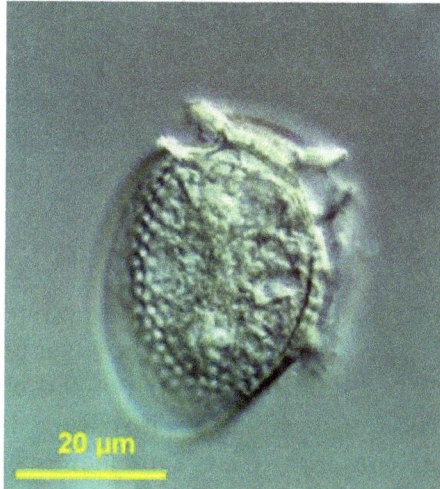

Dinophysis acuminata has chloroplasts taken from a cryptophyte

Cryptophyte Derived Dinophyte Chloroplast

Members of the genus *Dinophysis* have a phycobilin-containing chloroplast taken from a cryptophyte. However, the cryptophyte is not an endosymbiont—only the chloroplast seems to have been taken, and the chloroplast has been stripped of its nucleomorph and outermost two membranes, leaving just a two-membraned chloroplast. Cryptophyte chloroplasts require their nucleomorph to maintain themselves, and *Dinophysis* species grown in cell culture alone cannot survive, so it is possible (but not confirmed) that the *Dinophysis* chloroplast is a kleptoplast—if so, *Dinophysis* chloroplasts wear out and *Dinophysis* species must continually engulf cryptophytes to obtain new chloroplasts to replace the old ones.

Diatom Derived Dinophyte Chloroplasts

Some dinophytes, like *Kryptoperidinium* and *Durinskia* have a diatom (heterokontophyte) derived chloroplast. These chloroplasts are bounded by up to *five* membranes, (depending on whether you count the entire diatom endosymbiont as the chloroplast, or just the red algal derived chloroplast inside it). The diatom endosymbiont has been reduced relatively little—it still retains its original mitochondria, and has endoplasmic reticulum, ribosomes, a nucleus, and of course, red algal derived chloroplasts—practically a complete cell, all inside the host's endoplasmic reticulum lumen. However the diatom endosymbiont can't store its own food—its starch is found in granules in the dinophyte host's cytoplasm instead. The diatom endosymbiont's nucleus is present, but it probably can't be called a nucleomorph because it shows no sign of genome reduction, and might have even been *expanded*. Diatoms have been engulfed by dinoflagellates at least three times.

The diatom endosymbiont is bounded by a single membrane, inside it are chloroplasts with four membranes. Like the diatom endosymbiont's diatom ancestor, the chloroplasts have triplet thylakoids and pyrenoids.

In some of these genera, the diatom endosymbiont's chloroplasts aren't the only chloroplasts in the dinophyte. The original three-membraned peridinin chloroplast is still around, converted to an eyespot.

Prasinophyte (Green Algal) Derived Dinophyte Chloroplast

Lepidodinium viride and its close relatives are dinophytes that lost their original peridinin chloroplast and replaced it with a green algal derived chloroplast (more specifically, a prasinophyte).*Lepidodinium* is the only dinophyte that has a chloroplast that's not from the rhodoplast lineage. The chloroplast is surrounded by two membranes and has no nucleomorph—all the nucleomorph genes have been transferred to the dinophyte nucleus. The endosymbiotic event that led to this chloroplast was serial secondary endosymbiosis rather than tertiary endosymbiosis—the endosymbiont was a green alga containing a primary chloroplast (making a secondary chloroplast).

Chromatophores

While most chloroplasts originate from that first set of endosymbiotic events, *Paulinella chromatophora* is an exception that acquired a photosynthetic cyanobacterial endosymbiont more recently. It is not clear whether that symbiont is closely related to the ancestral chloroplast of other eukaryotes. Being in the early stages of endosymbiosis, *Paulinella chromatophora* can offer some insights into how chloroplasts evolved.*Paulinella* cells contain one or two sausage shaped blue-green photosynthesizing structures called chromatophores, descended from the cyanobacterium *Synechococcus*. Chromatophores cannot survive outside their host. Chromatophore DNA is about a million base pairs long, containing around 850 protein encoding genes—far less than the three million base pair *Synechococcus* genome, but much larger than the approximately 150,000 base pair genome of the more assimilated chloroplast. Chromatophores have transferred much less of their DNA to the nucleus of their host. About 0.3–0.8% of the nuclear DNA in *Paulinella* is from the chromatophore, compared with 11–14% from the chloroplast in plants.

Kleptoplastidy

In some groups of mixotrophicprotists, like some dinoflagellates, chloroplasts are separated from a captured alga or diatom and used temporarily. These klepto chloroplasts may only have a lifetime of a few days and are then replaced.

Chloroplast DNA

Chloroplasts have their own DNA, often abbreviated as ctDNA, or cpDNA. It is also known as the plastome. Its existence was first proved in 1962, and first sequenced in 1986—when two Japanese research teams sequenced the chloroplast DNA of liverwort and tobacco. Since then, hundreds of chloroplast DNAs from various species have been sequenced, but they're mostly those of land plants and green algae—glaucophytes, red algae, and other algal groups are extremely underrepresented, potentially introducing some bias in views of "typical" chloroplast DNA structure and content.

Molecular Structure

With few exceptions, most chloroplasts have their entire chloroplast genome combined into a single large circular DNA molecule, typically 120,000–170,000 base pairs long. They can have a contour length of around 30–60 micrometers, and have a mass of about 80–130 million daltons.

While usually thought of as a circular molecule, there is some evidence that chloroplast DNA molecules more often take on a linear shape.

Inverted Repeats

Many chloroplast DNAs contain two *inverted repeats*, which separate a long single copy section (LSC) from a short single copy section (SSC). While a given pair of inverted repeats are rarely completely identical, they are always very similar to each other, apparently resulting from concerted evolution.

The inverted repeats vary wildly in length, ranging from 4,000 to 25,000 base pairs long each and containing as few as four or as many as over 150 genes. Inverted repeats in plants tend to be at the upper end of this range, each being 20,000–25,000 base pairs long.

The inverted repeat regions are highly conserved among land plants, and accumulate few mutations. Similar inverted repeats exist in the genomes of cyanobacteria and the other two chloroplast lineages (glaucophyta and rhodophyceae), suggesting that they predate the chloroplast, though some chloroplast DNAs have since lost or flipped the inverted repeats (making them direct repeats). It is possible that the inverted repeats help stabilize the rest of the chloroplast genome, as chloroplast DNAs which have lost some of the inverted repeat segments tend to get rearranged more.

Nucleoids

New chloroplasts may contain up to 100 copies of their DNA, though the number of chloroplast DNA copies decreases to about 15–20 as the chloroplasts age. They are usually packed into nucleoids, which can contain several identical chloroplast DNA rings. Many nucleoids can be found in each chloroplast. In primitive red algae, the chloroplast DNA nucleoids are clustered in the center of the chloroplast, while in green plants and green algae, the nucleoids are dispersed throughout the stroma.

Though chloroplast DNA is not associated with true histones, in red algae, similar proteins that tightly pack each chloroplast DNA ring into a nucleoid have been found.

DNA Replication

The Leading Model of cpDNA Replication

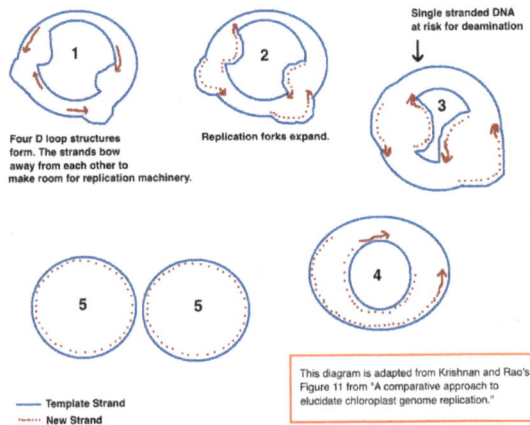

Chloroplast DNA replication via multiple D loop mechanisms. Adapted from Krishnan NM, Rao BJ's paper "A comparative approach to elucidate chloroplast genome replication."

The mechanism for chloroplast DNA (cpDNA) replication has not been conclusively determined, but two main models have been proposed. Scientists have attempted to observe chloroplast replication via electron microscopy since the 1970s. The results of the microscopy experiments led to the idea that chloroplast DNA replicates using a double displacement loop (D-loop). As the D-loop moves through the circular DNA, it adopts a theta intermediary form, also known as a Cairns replication intermediate, and completes replication with a rolling circle mechanism. Transcription starts at specific points of origin. Multiple replication forks open up, allowing replication machinery to transcribe the DNA. As replication continues, the forks grow and eventually converge. The new cpDNA structures separate, creating daughter cpDNA chromosomes.

In addition to the early microscopy experiments, this model is also supported by the amounts of deamination seen in cpDNA. Deamination occurs when an amino group is lost and is a mutation that often results in base changes. When adenine is deaminated, it becomes hypoxanthine. Hypoxanthine can bind to cytosine, and when the XC base pair is replicated, it becomes a GC (thus, an A → G base change).

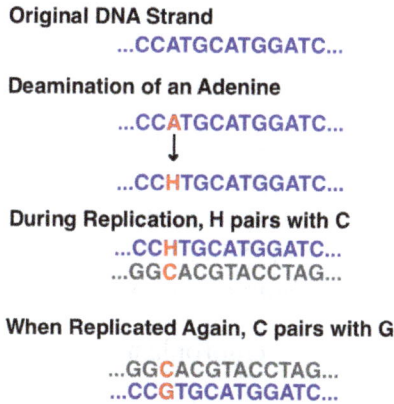

Original DNA Strand
...CCATGCATGGATC...

Deamination of an Adenine
...CCATGCATGGATC...
↓
...CCHTGCATGGATC...

During Replication, H pairs with C
...CCHTGCATGGATC...
...GGCACGTACCTAG...

When Replicated Again, C pairs with G
...GGCACGTACCTAG...
...CCGTGCATGGATC...

Over time, base changes in the DNA sequence can arise from deamination mutations. When adenine is deaminated, it becomes hypoxanthine, which can pair with cytosine. During replication, the cytosine will pair with guanine, causing an A --> G base change.

Deamination

In cpDNA, there are several A → G deamination gradients. DNA becomes susceptible to deamination events when it is single stranded. When replication forks form, the strand not being copied is single stranded, and thus at risk for A → G deamination. Therefore, gradients in deamination indicate that replication forks were most likely present and the direction that they initially opened (the highest gradient is most likely nearest the start site because it was single stranded for the longest amount of time). This mechanism is still the leading theory today; however, a second theory suggests that most cpDNA is actually linear and replicates through homologous recombination. It further contends that only a minority of the genetic material is kept in circular chromosomes while the rest is in branched, linear, or other complex structures.

Alternative Model of Replication

One of competing model for cpDNA replication asserts that most cpDNA is linear and participates in homologous recombination and replication structures similar to bacteriophage T4. It has been established that some plants have linear cpDNA, such as maize, and that more species still contain complex structures that scientists do not yet understand. When the original experiments on cpDNA were performed, scientists did notice linear structures; however, they attributed these linear forms to broken circles. If the branched and complex structures seen in cpDNA experiments are real and not artifacts of concatenated circular DNA or broken circles, then a D-loop mechanism of replication is insufficient to explain how those structures would replicate. At the same time, homologous recombination does not expand the multiple A --> G gradients seen in plastomes. Because of the failure to explain the deamination gradient as well as the

numerous plant species that have been shown to have circular cpDNA, the predomi-
nant theory continues to hold that most cpDNA is circular and most likely replicates via
a D loop mechanism.

Gene Content and Protein Synthesis

The chloroplast genome most commonly includes around 100 genes that code for a
variety of things, mostly to do with the protein pipeline and photosynthesis. As in prokaryotes, genes in chloroplast DNA are organized into operons. Interestingly though,
unlike prokaryotic DNA molecules, chloroplast DNA molecules contain introns (plant
mitochondrial DNAs do too, but not human mtDNAs).

Among land plants, the contents of the chloroplast genome are fairly similar.

Chloroplast Genome Reduction and Gene Transfer

Over time, many parts of the chloroplast genome were transferred to the nuclear
genome of the host, a process called *endosymbiotic gene transfer*. As a result, the
chloroplast genome is heavily reduced compared to that of free-living cyanobacteria.
Chloroplasts may contain 60–100 genes whereas cyanobacteria often have more than
1500 genes in their genome.Recently, a plastid without a genome was found, demon-
strating chloroplasts can lose their genome during endosymbiotic the gene transfer
process.

Endosymbiotic gene transfer is how we know about the lost chloroplasts in many chromalveolate lineages. Even if a chloroplast is eventually lost, the genes it donated to the
former host's nucleus persist, providing evidence for the lost chloroplast's existence.
For example, while diatoms (a heterokontophyte) now have a red algal derived chloroplast, the presence of many green algal genes in the diatom nucleus provide evidence
that the diatom ancestor (probably the ancestor of all chromalveolates too) had a green
algal derived chloroplast at some point, which was subsequently replaced by the red
chloroplast.

In land plants, some 11–14% of the DNA in their nuclei can be traced back to the chloroplast, up to 18% in *Arabidopsis*, corresponding to about 4,500 protein-coding genes.
There have been a few recent transfers of genes from the chloroplast DNA to the nuclear genome in land plants.

Of the approximately 3000 proteins found in chloroplasts, some 95% of them are encoded by nuclear genes. Many of the chloroplast's protein complexes consist of subunits from both the chloroplast genome and the host's nuclear genome. As a result,
protein synthesis must be coordinated between the chloroplast and the nucleus. The
chloroplast is mostly under nuclear control, though chloroplasts can also give out signals regulating gene expression in the nucleus, called *retrograde signaling*.

Protein Synthesis

Protein synthesis within chloroplasts relies on two RNA polymerases. One is coded by the chloroplast DNA, the other is of nuclear origin. The two RNA polymerases may recognize and bind to different kinds of promoters within the chloroplast genome. The ribosomes in chloroplasts are similar to bacterial ribosomes.

Protein Targeting and Import

Because so many chloroplast genes have been moved to the nucleus, many proteins that would originally have been translated in the chloroplast are now synthesized in the cytoplasm of the plant cell. These proteins must be directed back to the chloroplast, and imported through at least two chloroplast membranes.

Curiously, around half of the protein products of transferred genes aren't even targeted back to the chloroplast. Many became exaptations, taking on new functions like participating in cell division, protein routing, and even disease resistance. A few chloroplast genes found new homes in the mitochondrial genome—most became nonfunctional pseudogenes, though a few tRNA genes still work in the mitochondrion. Some transferred chloroplast DNA protein products get directed to the secretory pathway though it should be noted that many secondary plastids are bounded by an outermost membrane derived from the host's cell membrane, and therefore topologically outside of the cell, because to reach the chloroplast from the cytosol, you have to cross the cell membrane, just like if you were headed for the extracellular space. In those cases, chloroplast-targeted proteins do initially travel along the secretory pathway.

Because the cell acquiring a chloroplast already had mitochondria (and peroxisomes, and a cell membrane for secretion), the new chloroplast host had to develop a unique protein targeting system to avoid having chloroplast proteins being sent to the wrong organelle.

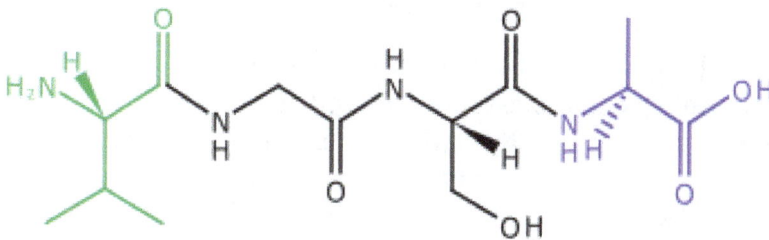

The two ends of a polypeptide are called the N-terminus, or *amino end*, and the C-terminus, or *carboxyl end*. This polypeptide has four amino acids linked together. At the left is the N-terminus, with its amino (H_2N) group in green. The blue C-terminus, with its carboxyl group (CO_2H) is at the right.

In most, but not all cases, nuclear-encoded chloroplast proteins are translated with a *cleavable transit peptide* that's added to the N-terminus of the protein precursor. Sometimes the transit sequence is found on the C-terminus of the protein, or within the functional part of the protein.

Transport Proteins and Membrane Translocons

After a chloroplast polypeptide is synthesized on a ribosome in the cytosol, an enzyme specific to chloroplast proteinsphosphorylates, or adds a phosphate group to many (but not all) of them in their transit sequences. Phosphorylation helps many proteins bind the polypeptide, keeping it from folding prematurely. This is important because it prevents chloroplast proteins from assuming their active form and carrying out their chloroplast functions in the wrong place—the cytosol. At the same time, they have to keep just enough shape so that they can be recognized by the chloroplast. These proteins also help the polypeptide get imported into the chloroplast.

From here, chloroplast proteins bound for the stroma must pass through two protein complexes—the TOC complex, or *translocon on the outer chloroplast membrane*, and the TIC translocon, or *translocon on the inner chloroplast membrane translocon*. Chloroplast polypeptide chains probably often travel through the two complexes at the same time, but the TIC complex can also retrieve preproteins lost in the intermembrane space.

Structure

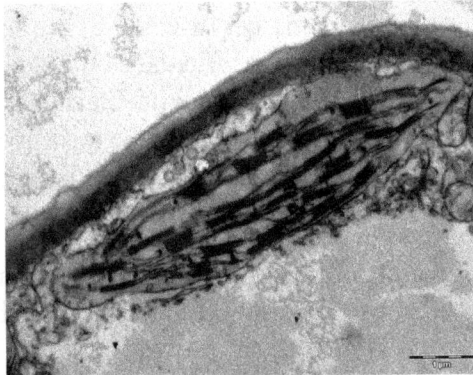

Transmission electron microscope image of a chloroplast. Grana of thylakoids and their connecting lamellae are clearly visible.

In land plants, chloroplasts are generally lens-shaped, 3–10 μm in diameter and 1–3 μm thick. Corn seedling chloroplasts are ≈20 μm³ in volume. Greater diversity in chloroplast shapes exists among the algae, which often contain a single chloroplast that can be shaped like a net (e.g., *Oedogonium*), a cup (e.g., *Chlamydomonas*), a ribbon-like spiral around the edges of the cell (e.g., *Spirogyra*), or slightly twisted bands at the cell edges (e.g., *Sirogonium*). Some algae have two chloroplasts in each cell; they are star-shaped in *Zygnema*, or may follow the shape of half the cell in orderDesmidiales. In some algae, the chloroplast takes up most of the cell, with pockets for the nucleus and other organelles, for example, some species of *Chlorella* have a cup-shaped chloroplast that occupies much of the cell.

All chloroplasts have at least three membrane systems—the outer chloroplast membrane, the inner chloroplast membrane, and the thylakoid system. Chloroplasts that are the product of secondary endosymbiosis may have additional membranes surrounding these three. Inside the outer and inner chloroplast membranes is the chloroplast stroma, a semi-gel-like fluid that makes up much of a chloroplast's volume, and in which the thylakoid system floats.

There are some common misconceptions about the outer and inner chloroplast membranes. The fact that chloroplasts are surrounded by a double membrane is often cited as evidence that they are the descendants of endosymbiotic cyanobacteria. This is often interpreted as meaning the outer chloroplast membrane is the product of the host's cell membrane infolding to form a vesicle to surround the ancestral cyanobacterium—which is not true—both chloroplast membranes are homologous to the cyanobacterium's original double membranes.

The chloroplast double membrane is also often compared to the mitochondrial double membrane. This is not a valid comparison—the inner mitochondria membrane is used to run proton pumps and carry out oxidative phosphorylation across to generate ATP energy. The only chloroplast structure that can consideredanalogous to it is the internal thylakoid system. Even so, in terms of "in-out", the direction of chloroplast H^+ ion flow is in the opposite direction compared to oxidative phosphorylation in mitochondria. In addition, in terms of function, the inner chloroplast membrane, which regulates metabolite passage and synthesizes some materials, has no counterpart in the mitochondrion.

Outer Chloroplast Membrane

The outer chloroplast membrane is a semi-porous membrane that small molecules and ions can easily diffuse across. However, it is not permeable to larger proteins, so chloroplast polypeptides being synthesized in the cell cytoplasm must be transported across the outer chloroplast membrane by the TOC complex, or *translocon on the outer chloroplast* membrane.

The chloroplast membranes sometimes protrude out into the cytoplasm, forming a stromule, or stroma-containing tubule. Stromules are very rare in chloroplasts, and are much more common in other plastids like chromoplasts and amyloplasts in petals and roots, respectively. They may exist to increase the chloroplast's surface area for cross-membrane transport, because they are often branched and tangled with the endoplasmic reticulum. When they were first observed in 1962, some plant biologists dismissed the structures as artifactual, claiming that stromules were just oddly shaped chloroplasts with constricted regions or dividing chloroplasts. However, there is a growing body of evidence that stromules are functional, integral features of plant cell plastids, not merely artifacts.

Intermembrane Space and Peptidoglycan Wall

Usually, a thin intermembrane space about 10–20 nanometers thick exists between the outer and inner chloroplast membranes.

Instead of an intermembrane space, glaucophyte algae have a peptidoglycan wall
between their inner and outer chloroplast membranes.

Glaucophyte algal chloroplasts have a peptidoglycan layer between the chloroplast
membranes. It corresponds to the peptidoglycan cell wall of their cyanobacterial ancestors, which is located between their two cell membranes. These chloroplasts are
called *muroplasts* (from Latin *"mura"*, meaning "wall"). Other chloroplasts have lost
the cyanobacterial wall, leaving an intermembrane space between the two chloroplast
envelope membranes.

Inner Chloroplast Membrane

The inner chloroplast membrane borders the stroma and regulates passage of materials
in and out of the chloroplast. After passing through the TOC complex in the outer chloroplast membrane, polypeptides must pass through the TIC complex*(translocon on
the inner chloroplast membrane)* which is located in the inner chloroplast membrane.

In addition to regulating the passage of materials, the inner chloroplast membrane is
where fatty acids, lipids, and carotenoids are synthesized.

Peripheral Reticulum

Some chloroplasts contain a structure called the chloroplast peripheral reticulum. It is often found in the chloroplasts of C_4 plants, though it has also been found in some C_3 angiosperms, and even some gymnosperms. The chloroplast peripheral reticulum consists of a
maze of membranous tubes and vesicles continuous with the inner chloroplast membrane
that extends into the internal stromal fluid of the chloroplast. Its purpose is thought to be
to increase the chloroplast's surface area for cross-membrane transport between its stroma and the cell cytoplasm. The small vesicles sometimes observed may serve as transport
vesicles to shuttle stuff between the thylakoids and intermembrane space.

Stroma

The protein-rich,alkaline,aqueous fluid within the inner chloroplast membrane and outside of the thylakoid space is called the stroma, which corresponds to the cytosol of the

original cyanobacterium. Nucleoids of chloroplast DNA, chloroplast ribosomes, the thylakoid system with plastoglobuli, starch granules, and many proteins can be found floating around in it. The Calvin cycle, which fixes CO_2 into sugar takes place in the stroma.

Chloroplast Ribosomes

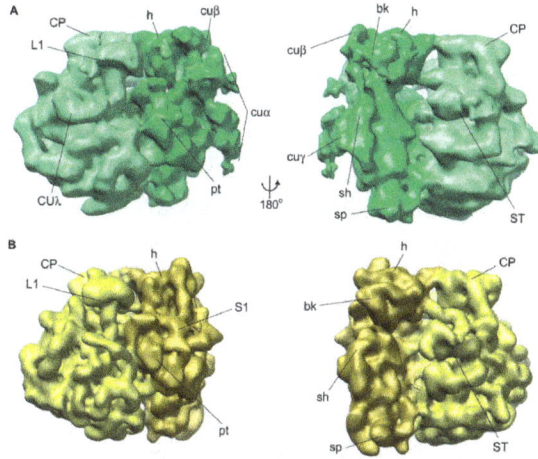

Chloroplast ribosomes Comparison of a chloroplast ribosome (green) and a bacterial ribosome (yellow). Important features common to both ribosomes and chloroplast-unique features are labeled.

Chloroplasts have their own ribosomes, which they use to synthesize a small fraction of their proteins. Chloroplast ribosomes are about two-thirds the size of cytoplasmic ribosomes (around 17 nm vs 25 nm). They take mRNAs transcribed from the chloroplast DNA and translate them into protein. While similar to bacterial ribosomes, chloroplast translation is more complex than in bacteria, so chloroplast ribosomes include some chloroplast-unique features. Small subunit ribosomal RNAs in several Chlorophyta and euglenid chloroplasts lack motifs for shine-dalgarno sequence recognition, which is considered essential for translation initiation in most chloroplasts and prokaryotes. Such loss is also rarely observed in other plastids and prokaryotes.

Plastoglobuli

Plastoglobuli (singular*plastoglobulus*, sometimes spelled *plastoglobule(s)*), are spherical bubbles of lipids and proteins about 45–60 nanometers across.They are surrounded by a lipid monolayer. Plastoglobuli are found in all chloroplasts, but become more common when the chloroplast is under oxidative stress, or when it ages and transitions into a gerontoplast. Plastoglobuli also exhibit a greater size variation under these conditions.They are also common in etioplasts, but decrease in number as the etioplasts mature into chloroplasts.

Plastoglubuli contain both structural proteins and enzymes involved in lipid synthesis and metabolism. They contain many types of lipids including plastoquinone, vitamin E, carotenoids and chlorophylls.

Plastoglobuli were once thought to be free-floating in the stroma, but it is now thought that they are permanently attached either to a thylakoid or to another plastoglobulus attached to a thylakoid, a configuration that allows a plastoglobulus to exchange its contents with the thylakoid network.In normal green chloroplasts, the vast majority of plastoglobuli occur singularly, attached directly to their parent thylakoid. In old or stressed chloroplasts, plastoglobuli tend to occur in linked groups or chains, still always anchored to a thylakoid.

Plastoglobuli form when a bubble appears between the layers of the lipid bilayer of the thylakoid membrane, or bud from existing plastoglubuli—though they never detach and float off into the stroma. Practically all plastoglobuli form on or near the highly curved edges of the thylakoid disks or sheets. They are also more common on stromal thylakoids than on granal ones.

Starch Granules

Starch granules are very common in chloroplasts, typically taking up 15% of the organelle's volume, though in some other plastids like amyloplasts, they can be big enough to distort the shape of the organelle. Starch granules are simply accumulations of starch in the stroma, and are not bounded by a membrane.

Starch granules appear and grow throughout the day, as the chloroplast synthesizes sugars, and are consumed at night to fuel respiration and continue sugar export into the phloem,though in mature chloroplasts, it is rare for a starch granule to be completely consumed or for a new granule to accumulate.

Starch granules vary in composition and location across different chloroplast lineages. In red algae, starch granules are found in the cytoplasm rather than in the chloroplast. In C_4 plants, mesophyll chloroplasts, which do not synthesize sugars, lack starch granules.

Rubisco

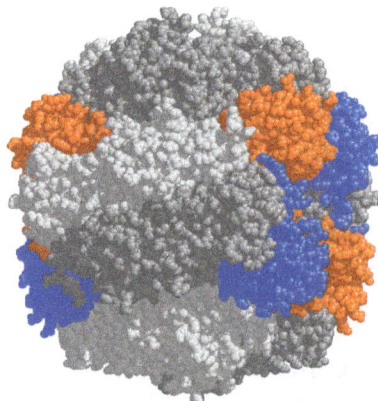

Rubisco, shown here in a space-filling model, is the main enzyme responsible for carbon fixation in chloroplasts.

The chloroplast stroma contains many proteins, though the most common and important is Rubisco, which is probably also the most abundant protein on the planet.Rubisco is the enzyme that fixes CO_2 into sugar molecules. In C_3 plants, rubisco is abundant in all chloroplasts, though in C_4 plants, it is confined to the bundle sheath chloroplasts, where the Calvin cycle is carried out in C_4 plants.

Pyrenoids

The chloroplasts of some hornworts and algae contain structures called pyrenoids. They are not found in higher plants. Pyrenoids are roughly spherical and highly refractive bodies which are a site of starch accumulation in plants that contain them. They consist of a matrix opaque to electrons, surrounded by two hemispherical starch plates. The starch is accumulated as the pyrenoids mature.In algae with carbon concentrating mechanisms, the enzyme rubisco is found in the pyrenoids. Starch can also accumulate around the pyrenoids when CO_2 is scarce. Pyrenoids can divide to form new pyrenoids, or be produced "de novo".

Thylakoid System

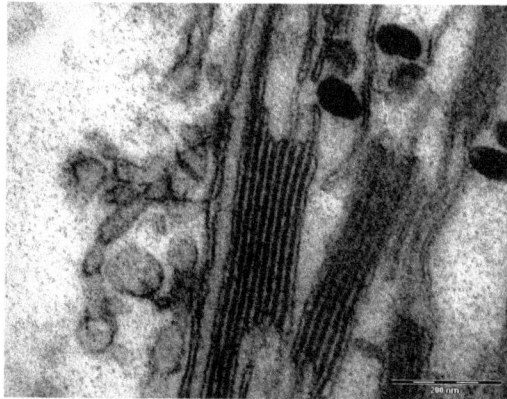

Transmission electron microscope image of some thylakoids arranged in grana stacks and lamellæ. Plastoglobuli (dark blobs) are also present.

Suspended within the chloroplast stroma is the thylakoid system, a highly dynamic collection of membranous sacks called thylakoids where chlorophyll is found and the light reactions of photosynthesis happen. In most vascular plant chloroplasts, the thylakoids are arranged in stacks called grana, though in certain C_4 plant chloroplasts and some algal chloroplasts, the thylakoids are free floating.

Granal Structure

Using a light microscope, it is just barely possible to see tiny green granules—which were named grana. With electron microscopy, it became possible to see the thylakoid system in more detail, revealing it to consist of stacks of flat thylakoids which made up the grana, and long interconnecting stromal thylakoids which linked different grana.

In the transmission electron microscope, thylakoid membranes appear as alternating light-and-dark bands, 8.5 nanometers thick.

For a long time, the three-dimensional structure of the thylakoid system has been unknown or disputed. One model has the granum as a stack of thylakoids linked by helical stromal thylakoids; the other has the granum as a single folded thylakoid connected in a "hub and spoke" way to other grana by stromal thylakoids. While the thylakoid system is still commonly depicted according to the folded thylakoid model, it was determined in 2011 that the stacked and helical thylakoids model is correct.

In the helical thylakoid model, grana consist of a stack of flattened circular granal thylakoids that resemble pancakes. Each granum can contain anywhere from two to a hundred thylakoids, though grana with 10–20 thylakoids are most common. Wrapped around the grana are helicoid stromal thylakoids, also known as frets or lamellar thylakoids. The helices ascend at an angle of 20–25°, connecting to each granal thylakoid at a bridge-like slit junction. The helicoids may extend as large sheets that link multiple grana, or narrow to tube-like bridges between grana.While different parts of the thylakoid system contain different membrane proteins, the thylakoid membranes are continuous and the thylakoid space they enclose form a single continuous labyrinth.

Thylakoids

Thylakoids (sometimes spelled *thylakoïds*), are small interconnected sacks which contain the membranes that the light reactions of photosynthesis take place on. The word *thylakoid* comes from the Greek word *thylakos* which means "sack".

Embedded in the thylakoid membranes are important protein complexes which carry out the light reactions of photosynthesis. Photosystem II and photosystem I contain light-harvesting complexes with chlorophyll and carotenoids that absorb light energy and use it to energize electrons. Molecules in the thylakoid membrane use the energized electrons to pump hydrogen ions into the thylakoid space, decreasing the pH and turning it acidic. ATP synthase is a large protein complex that harnesses the concentration gradient of the hydrogen ions in the thylakoid space to generate ATP energy as the hydrogen ions flow back out into the stroma—much like a dam turbine.

There are two types of thylakoids—granal thylakoids, which are arranged in grana, and stromal thylakoids, which are in contact with the stroma. Granal thylakoids are pancake-shaped circular disks about 300–600 nanometers in diameter. Stromal thylakoids are helicoid sheets that spiral around grana.The flat tops and bottoms of granal thylakoids contain only the relatively flat photosystem II protein complex. This allows them to stack tightly, forming grana with many layers of tightly appressed membrane, called granal membrane, increasing stability and surface area for light capture.

In contrast, photosystem I and ATP synthase are large protein complexes which jut out into the stroma. They can't fit in the appressed granal membranes, and so are found in the stromal thylakoid membrane—the edges of the granal thylakoid disks and the stromal thylakoids. These large protein complexes may act as spacers between the sheets of stromal thylakoids.

The number of thylakoids and the total thylakoid area of a chloroplast is influenced by light exposure. Shaded chloroplasts contain larger and more grana with more thylakoid membrane area than chloroplasts exposed to bright light, which have smaller and fewer grana and less thylakoid area. Thylakoid extent can change within minutes of light exposure or removal.

Pigments and Chloroplast Colors

Inside the photosystems embedded in chloroplast thylakoid membranes are various photosynthetic pigments, which absorb and transfer light energy. The types of pigments found are different in various groups of chloroplasts, and are responsible for a wide variety of chloroplast colorations.

Chlorophylls

Chlorophyll *a* is found in all chloroplasts, as well as their cyanobacterial ancestors. Chlorophyll *a* is a blue-greenpigment partially responsible for giving most cyanobacteria and chloroplasts their color. Other forms of chlorophyll exist, such as the accessory pigmentschlorophyll *b*, chlorophyll *c*, chlorophyll *d*, and chlorophyll *f*.

Chlorophyll *b* is an olive green pigment found only in the chloroplasts of plants, green algae, any secondary chloroplasts obtained through the secondary endosymbiosis of a green alga, and a few cyanobacteria. It is the chlorophylls *a* and *b* together that make most plant and green algal chloroplasts green.

Chlorophyll *c* is mainly found in secondary endosymbiotic chloroplasts that originated from a red alga, although it is not found in chloroplasts of red algae themselves. Chlorophyll *c* is also found in some green algae and cyanobacteria.

Chlorophylls *d* and *f* are pigments found only in some cyanobacteria.

Carotenoids

In addition to chlorophylls, another group of yellow–orange pigments called carotenoids are also found in the photosystems. There are about thirty photosynthetic carotenoids. They help transfer and dissipate excess energy, and their bright colors sometimes override the chlorophyll green, like during the fall, when the leaves of some land plants change color.β-carotene is a bright red-orange carotenoid found in nearly all chloroplasts, like chlorophyll *a*.Xanthophylls, especially the orange-red zeaxanthin, are also common.Many other forms of carotenoids exist that are only found in certain groups of chloroplasts.

Delesseria sanguinea, a red alga, has chloroplasts that contain red pigments like phycoerytherin that mask their blue-green chlorophyll *a*.

Phycobilins

Phycobilins are a third group of pigments found in cyanobacteria, and glaucophyte, red algal, and cryptophyte chloroplasts. Phycobilins come in all colors, though phycoerytherin is one of the pigments that makes many red algae red. Phycobilins often organize into relatively large protein complexes about 40 nanometers across called phycobilisomes. Like photosystem I and ATP synthase, phycobilisomes jut into the stroma, preventing thylakoid stacking in red algal chloroplasts.Cryptophyte chloroplasts and some cyanobacteria don't have their phycobilin pigments organized into phycobilisomes, and keep them in their thylakoid space instead.

Specialized Chloroplasts in C4 Plants

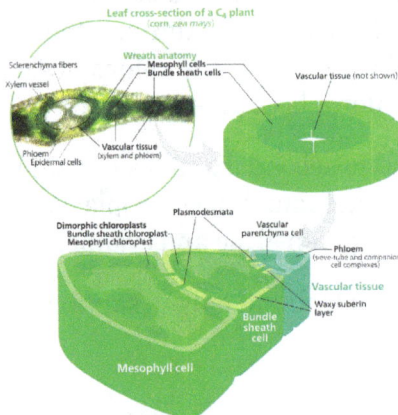

Many C$_4$ plants have their mesophyll cells and bundle sheath cells arranged radially around their leaf veins. The two types of cells contain different types of chloroplasts specialized for a particular part of photosynthesis.

To fix carbon dioxide into sugar molecules in the process of photosynthesis, chloroplasts use an enzyme called rubisco. Rubisco has a problem—it has trouble distinguishing between carbon dioxide and oxygen, so at high oxygen concentrations, rubisco starts accidentally adding oxygen to sugar precursors. This has the end result of ATP energy being wasted and CO$_2$ being released, all with no sugar being produced. This is a big problem, since O_2 is produced by the initial light reactions of photosynthesis, causing issues down the line in the Calvin cycle which uses rubisco.

C_4 plants evolved a way to solve this—by spatially separating the light reactions and the Calvin cycle. The light reactions, which store light energy in ATP and NADPH, are done in the mesophyll cells of a C_4 leaf. The Calvin cycle, which uses the stored energy to make sugar using rubisco, is done in the bundle sheath cells, a layer of cells surrounding a vein in a leaf.

As a result, chloroplasts in C_4 mesophyll cells and bundle sheath cells are specialized for each stage of photosynthesis. In mesophyll cells, chloroplasts are specialized for the light reactions, so they lack rubisco, and have normal grana and thylakoids, which they use to make ATP and NADPH, as well as oxygen. They store CO_2 in a four-carbon compound, which is why the process is called C_4 *photosynthesis*. The four-carbon compound is then transported to the bundle sheath chloroplasts, where it drops off CO_2 and returns to the mesophyll. Bundle sheath chloroplasts do not carry out the light reactions, preventing oxygen from building up in them and disrupting rubisco activity. Because of this, they lack thylakoids organized into grana stacks—though bundle sheath chloroplasts still have free-floating thylakoids in the stroma where they still carry out cyclic electron flow, a light-driven method of synthesizing ATP to power the Calvin cycle without generating oxygen. They lack photosystem II, and only have photosystem I—the only protein complex needed for cyclic electron flow. Because the job of bundle sheath chloroplasts is to carry out the Calvin cycle and make sugar, they often contain large starch grains.

Both types of chloroplast contain large amounts of chloroplast peripheral reticulum, which they use to get more surface area to transport stuff in and out of them. Mesophyll chloroplasts have a little more peripheral reticulum than bundle sheath chloroplasts.

Location

Distribution in a Plant

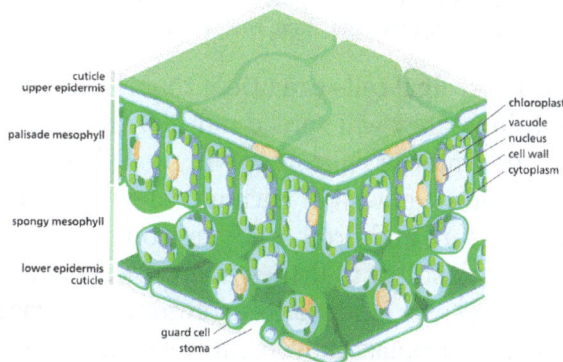

A cross section of a leaf, showing chloroplasts in its mesophyll cells. Stomal guard cells also have chloroplasts, though much fewer than mesophyll cells.

Not all cells in a multicellular plant contain chloroplasts. All green parts of a plant contain chloroplasts—the chloroplasts, or more specifically, the chlorophyll in them are what make the photosynthetic parts of a plant green. The plant cells which contain chloroplasts are usually parenchyma cells, though chloroplasts can also be found in collenchyma tissue.A plant cell which contains chloroplasts is known as a chlorenchyma cell. A typical chlorenchyma cell of a land plant contains about 10 to 100 chloroplasts.

In some plants such as cacti, chloroplasts are found in the stems, though in most plants, chloroplasts are concentrated in the leaves. One square millimeter of leaf tissue can contain half a million chloroplasts. Within a leaf, chloroplasts are mainly found in the mesophyll layers of a leaf, and the guard cells of stomata. Palisade mesophyll cells can contain 30–70 chloroplasts per cell, while stomatal guard cells contain only around 8–15 per cell, as well as much less chlorophyll. Chloroplasts can also be found in the bundle sheath cells of a leaf, especially in C_4 plants, which carry out the Calvin cycle in their bundle sheath cells. They are often absent from the epidermis of a leaf.

Cellular Location

Chloroplast Movement

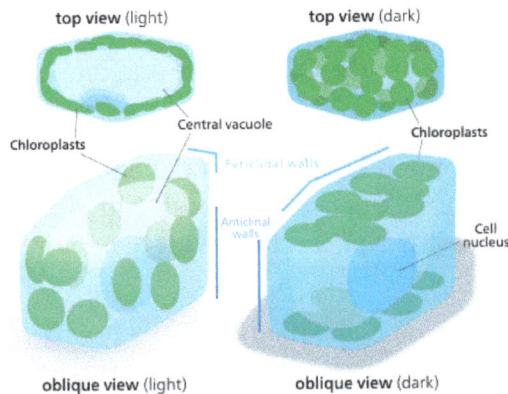

When chloroplasts are exposed to direct sunlight, they stack along the anticlinal cell walls to minimize exposure. In the dark they spread out in sheets along the periclinal walls to maximize light absorption.

The chloroplasts of plant and algal cells can orient themselves to best suit the available light. In low-light conditions, they will spread out in a sheet—maximizing the surface area to absorb light. Under intense light, they will seek shelter by aligning in vertical columns along the plant cell's cell wall or turning sideways so that light strikes them edge-on. This reduces exposure and protects them from photooxidative damage. This ability to distribute chloroplasts so that they can take shelter behind each other or spread out may be the reason why land plants evolved to have many small chloroplasts instead of a few big ones. Chloroplast movement is considered one of the most closely regulated stimulus-response systems that can be found in plants.Mitochondria have also been observed to follow chloroplasts as they move.

In higher plants, chloroplast movement is run by phototropins, blue light photoreceptors also responsible for plant phototropism. In some algae, mosses, ferns, and flowering plants, chloroplast movement is influenced by red light in addition to blue light, though very long red wavelengths inhibit movement rather than speeding it up. Blue light generally causes chloroplasts to seek shelter, while red light draws them out to maximize light absorption.

Studies of *Vallisneria gigantea*, an aquatic flowering plant, have shown that chloroplasts can get moving within five minutes of light exposure, though they don't initially show any net directionality. They may move along microfilament tracks, and the fact that the microfilament mesh changes shape to form a honeycomb structure surrounding the chloroplasts after they have moved suggests that microfilaments may help to anchor chloroplasts in place.

Function and Chemistry

Guard Cell Chloroplasts

Unlike most epidermal cells, the guard cells of plant stomata contain relatively well-developed chloroplasts. However, exactly what they do is controversial.

Plant Innate Immunity

Plants lack specialized immune cells—all plant cells participate in the plant immune response. Chloroplasts, along with the nucleus, cell membrane, and endoplasmic reticulum, are key players in pathogen defense. Due to its role in a plant cell's immune response, pathogens frequently target the chloroplast.

Plants have two main immune responses—the hypersensitive response, in which infected cells seal themselves off and undergo programmed cell death, and systemic acquired resistance, where infected cells release signals warning the rest of the plant of a pathogen's presence. Chloroplasts stimulate both responses by purposely damaging their photosynthetic system, producing reactive oxygen species. High levels of reactive oxygen species will cause the hypersensitive response. The reactive oxygen species also directly kill any pathogens within the cell. Lower levels of reactive oxygen species initiate systemic acquired resistance, triggering defense-molecule production in the rest of the plant.

In some plants, chloroplasts are known to move closer to the infection site and the nucleus during an infection.

Chloroplasts can serve as cellular sensors. After detecting stress in a cell, which might be due to a pathogen, chloroplasts begin producing molecules like salicylic acid, jasmonic acid, nitric oxide and reactive oxygen species which can serve as defense-signals. As cellular signals, reactive oxygen species are unstable molecules, so they probably don't leave the chloroplast, but instead pass on their signal to an unknown second messenger molecule. All these molecules initiate retrograde signaling—signals from the chloroplast that regulate gene expression in the nucleus.

In addition to defense signaling, chloroplasts, with the help of the peroxisomes, help synthesize an important defense molecule, jasmonate. Chloroplasts synthesize all the fatty acids in a plant cell—linoleic acid, a fatty acid, is a precursor to jasmonate.

Photosynthesis

One of the main functions of the chloroplast is its role in photosynthesis, the process by which light is transformed into chemical energy, to subsequently produce food in the form of sugars. Water (H_2O) and carbon dioxide (CO_2) are used in photosynthesis, and sugar and oxygen (O_2) is made, using light energy. Photosynthesis is divided into two stages—the light reactions, where water is split to produce oxygen, and the dark reactions, or Calvin cycle, which builds sugar molecules from carbon dioxide. The two phases are linked by the energy carriersadenosine triphosphate (ATP) and nicotinamide adenine dinucleotide phosphate (NADP$^+$).

Light Reactions

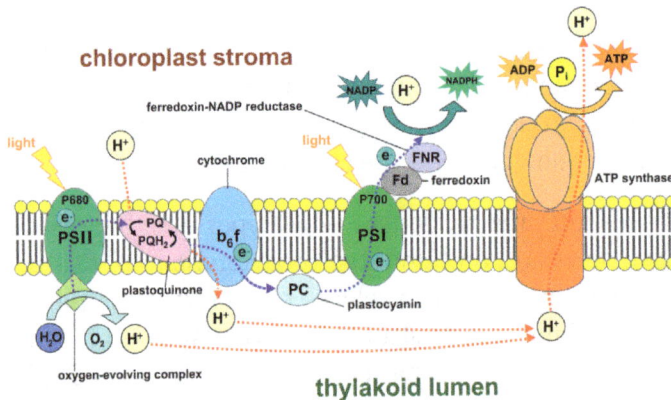

The light reactions of photosynthesis take place across the thylakoid membranes

The light reactions take place on the thylakoid membranes. They take light energy and store it in NADPH, a form of NADP$^+$, and ATP to fuel the dark reactions.

Energy Carriers

ATP is the phosphorylated version of adenosine diphosphate (ADP), which stores energy in a cell and powers most cellular activities. ATP is the energized form, while ADP is the (partially) depleted form. NADP$^+$ is an electron carrier which ferries high energy electrons. In the light reactions, it gets reduced, meaning it picks up electrons, becoming NADPH.

Photophosphorylation

Like mitochondria, chloroplasts use the potential energy stored in an H^+, or hydrogen ion gradient to generate ATP energy. The two photosystems capture light energy to energize electrons taken from water, and release them down an electron transport chain. The molecules between the photosystems harness the electrons' energy to pump hydrogen ions into the thylakoid space, creating a concentration gradient, with more hydrogen ions (up to a thousand times as many) inside the thylakoid system than in the stroma. The hydrogen ions in the thylakoid space then diffuse back down their concentration gradient, flowing back out into the stroma through ATP synthase. ATP synthase uses the energy from the flowing hydrogen ions to phosphorylateadenosine diphosphate into adenosine triphosphate, or ATP. Because chloroplast ATP synthase projects out into the stroma, the ATP is synthesized there, in position to be used in the dark reactions.

NADP+Reduction

Electrons are often removed from the electron transport chains to charge $NADP^+$ with electrons, reducing it to NADPH. Like ATP synthase, ferredoxin-$NADP^+$ reductase, the enzyme that reduces $NADP^+$, releases the NADPH it makes into the stroma, right where it is needed for the dark reactions.

Because $NADP^+$ reduction removes electrons from the electron transport chains, they must be replaced—the job of photosystem II, which splits water molecules (H_2O) to obtain the electrons from its hydrogen atoms.

Cyclic Photophosphorylation

While photosystem IIphotolyzes water to obtain and energize new electrons, photosystem I simply reenergizes depleted electrons at the end of an electron transport chain. Normally, the reenergized electrons are taken by $NADP^+$, though sometimes they can flow back down more H^+-pumping electron transport chains to transport more hydrogen ions into the thylakoid space to generate more ATP. This is termed cyclic photophosphorylation because the electrons are recycled. Cyclic photophosphorylation is common in C_4 plants, which need more ATP than NADPH.

Dark Reactions

The Calvin cycle, also known as the dark reactions, is a series of biochemical reactions that fixes CO_2 into G3P sugar molecules and uses the energy and electrons from the ATP and NADPH made in the light reactions. The Calvin cycle takes place in the stroma of the chloroplast.

While named *"the dark reactions"*, in most plants, they take place in the light, since the dark reactions are dependent on the products of the light reactions.

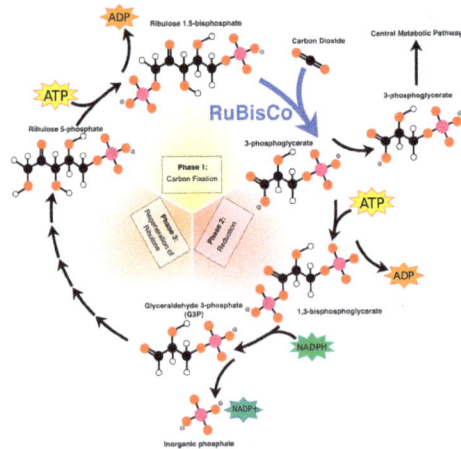

The Calvin cycle incorporates carbon dioxide
into sugar molecules.

Carbon Fixation and G3P Synthesis

The Calvin cycle starts by using the enzyme Rubisco to fix CO_2 into five-carbon Ribulose bisphosphate (RuBP) molecules. The result is unstable six-carbon molecules that immediately break down into three-carbon molecules called 3-phosphoglyceric acid, or 3-PGA. The ATP and NADPH made in the light reactions is used to convert the 3-PGA into glyceraldehyde-3-phosphate, or G3P sugar molecules. Most of the G3P molecules are recycled back into RuBP using energy from more ATP, but one out of every six produced leaves the cycle—the end product of the dark reactions.

Sugars and Starches

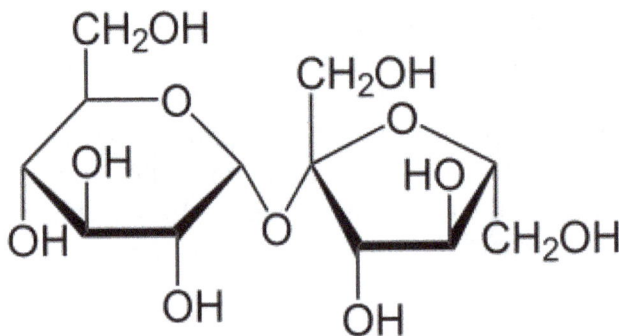

Sucrose is made up of a glucose monomer (left), and a fructose monomer (right)

Glyceraldehyde-3-phosphate can double up to form larger sugar molecules like glucose and fructose. These molecules are processed, and from them, the still larger sucrose, a disaccharide commonly known as table sugar, is made, though this process takes place outside of the chloroplast, in the cytoplasm.

Alternatively, glucose monomers in the chloroplast can be linked together to make starch, which accumulates into the starch grains found in the chloroplast.Under conditions such as high atmospheric CO_2 concentrations, these starch grains may grow very large, distorting the grana and thylakoids. The starch granules displace the thylakoids, but leave them intact. Waterlogged roots can also cause starch buildup in the chloroplasts, possibly due to less sucrose being exported out of the chloroplast (or more accurately, the plant cell). This depletes a plant's free phosphate supply, which indirectly stimulates chloroplast starch synthesis. While linked to low photosynthesis rates, the starch grains themselves may not necessarily interfere significantly with the efficiency of photosynthesis, and might simply be a side effect of another photosynthesis-depressing factor.

Photorespiration

Photorespiration can occur when the oxygen concentration is too high. Rubisco cannot distinguish between oxygen and carbon dioxide very well, so it can accidentally add O_2 instead of CO_2 to RuBP. This process reduces the efficiency of photosynthesis—it consumes ATP and oxygen, releases CO_2, and produces no sugar. It can waste up to half the carbon fixed by the Calvin cycle. Several mechanisms have evolved in different lineages that raise the carbon dioxide concentration relative to oxygen within the chloroplast, increasing the efficiency of photosynthesis. These mechanisms are called carbon dioxide concentrating mechanisms, or CCMs. These include Crassulacean acid metabolism, C_4 carbon fixation, and pyrenoids. Chloroplasts in C_4 plants are notable as they exhibit a distinct chloroplast dimorphism.

pH

Because of the H^+ gradient across the thylakoid membrane, the interior of the thylakoid is acidic, with a pH around 4, while the stroma is slightly basic, with a pH of around 8.The optimal stroma pH for the Calvin cycle is 8.1, with the reaction nearly stopping when the pH falls below 7.3.

CO_2 in water can form carbonic acid, which can disturb the pH of isolated chloroplasts, interfering with photosynthesis, even though CO_2 is used in photosynthesis. However, chloroplasts in living plant cells are not affected by this as much.

Chloroplasts can pump K^+ and H^+ ions in and out of themselves using a poorly understood light-driven transport system.

In the presence of light, the pH of the thylakoid lumen can drop up to 1.5 pH units, while the pH of the stroma can rise by nearly one pH unit.

Amino Acid Synthesis

Chloroplasts alone make almost all of a plant cell's amino acids in their stroma except the sulfur-containing ones like cysteine and methionine. Cysteine is made in the chloroplast (the proplastid too) but it is also synthesized in the cytosol and mitochondria, probably because it has trouble crossing membranes to get to where it is needed.The chloroplast is known to make the precursors to methionine but it is unclear whether the organelle carries out the last leg of the pathway or if it happens in the cytosol.

Other Nitrogen Compounds

Chloroplasts make all of a cell's purines and pyrimidines—the nitrogenous bases found in DNA and RNA. They also convert nitrite (NO_2^-) into ammonia (NH_3) which supplies the plant with nitrogen to make its amino acids and nucleotides.

Other Chemical Products

Chloroplasts are the site of complex lipid metabolism.

Differentiation, Replication, and Inheritance

Chloroplasts are a special type of a plant cell organelle called a plastid, though the two terms are sometimes used interchangeably. There are many other types of plastids, which carry out various functions. All chloroplasts in a plant are descended from undifferentiated proplastids found in the zygote, or fertilized egg. Proplastids are commonly found in an adult plant's apical meristems. Chloroplasts do not normally develop from proplastids in root tip meristems—instead, the formation of starch-storing amyloplasts is more common.

In shoots, proplastids from shoot apical meristems can gradually develop into chloroplasts in photosynthetic leaf tissues as the leaf matures, if exposed to the required light. This process involves invaginations of the inner plastid membrane, forming sheets of membrane that project into the internal stroma. These membrane sheets then fold to form thylakoids and grana.

If angiosperm shoots are not exposed to the required light for chloroplast formation, proplastids may develop into an etioplast stage before becoming chloroplasts. An etioplast is a plastid that lacks chlorophyll, and has inner membrane invaginations that form a lattice of tubes in their stroma, called a prolamellar body. While etioplasts lack chlorophyll, they have a yellow chlorophyll precursor stocked. Within a few minutes of light exposure, the prolamellar body begins to reorganize into stacks of thylakoids, and chlorophyll starts to be produced. This process, where the etioplast becomes a chloroplast, takes several hours.Gymnosperms do not require light to form chloroplasts.

Light, however, does not guarantee that a proplastid will develop into a chloroplast. Whether a proplastid develops into a chloroplast some other kind of plastid is mostly controlled by the nucleus and is largely influenced by the kind of cell it resides in.

possible plastid interconversions

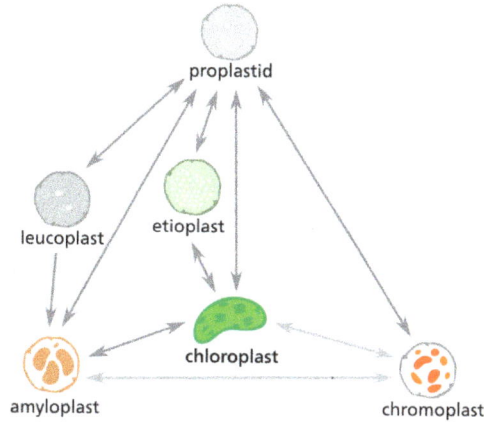

Many plastid interconversions are possible

Plastid Interconversion

Plastid differentiation is not permanent, in fact many interconversions are possible. Chloroplasts may be converted to chromoplasts, which are pigment-filled plastids responsible for the bright colors seen in flowers and ripe fruit. Starch storing amyloplasts can also be converted to chromoplasts, and it is possible for proplastids to develop straight into chromoplasts. Chromoplasts and amyloplasts can also become chloroplasts, like what happens when a carrot or a potato is illuminated. If a plant is injured, or something else causes a plant cell to revert to a meristematic state, chloroplasts and other plastids can turn back into proplastids. Chloroplast, amyloplast, chromoplast, proplast, etc., are not absolute states—intermediate forms are common.

Chloroplast Division

Most chloroplasts in a photosynthetic cell do not develop directly from proplastids or etioplasts. In fact, a typical shoot meristematic plant cell contains only 7–20 proplastids. These proplastids differentiate into chloroplasts, which divide to create the 30–70 chloroplasts found in a mature photosynthetic plant cell. If the cell divides, chloroplast division provides the additional chloroplasts to partition between the two daughter cells.

In single-celled algae, chloroplast division is the only way new chloroplasts are formed. There is no proplastid differentiation—when an algal cell divides, its chloroplast divides along with it, and each daughter cell receives a mature chloroplast.

Almost all chloroplasts in a cell divide, rather than a small group of rapidly dividing chloroplasts. Chloroplasts have no definite S-phase—their DNA replication is not synchronized or limited to that of their host cells. Much of what we know about chloroplast division comes from studying organisms like *Arabidopsis* and the red alga *Cyanidioschyzon merolæ*.

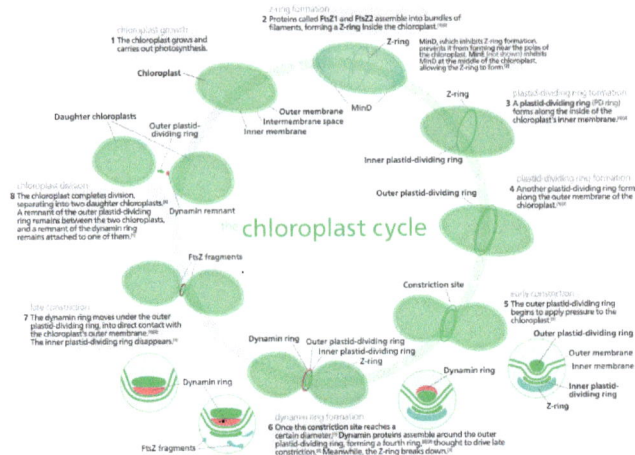

Most chloroplasts in plant cells, and all chloroplasts in algae arise from chloroplast division

The division process starts when the proteins FtsZ1 and FtsZ2 assemble into filaments, and with the help of a protein ARC6, form a structure called a Z-ring within the chloroplast's stroma. The Min system manages the placement of the Z-ring, ensuring that the chloroplast is cleaved more or less evenly. The protein MinD prevents FtsZ from linking up and forming filaments. Another protein ARC3 may also be involved, but it is not very well understood. These proteins are active at the poles of the chloroplast, preventing Z-ring formation there, but near the center of the chloroplast, MinE inhibits them, allowing the Z-ring to form.

Next, the two plastid-dividing rings, or PD rings form. The inner plastid-dividing ring is located in the inner side of the chloroplast's inner membrane, and is formed first. The outer plastid-dividing ring is found wrapped around the outer chloroplast membrane. It consists of filaments about 5 nanometers across, arranged in rows 6.4 nanometers apart, and shrinks to squeeze the chloroplast. This is when chloroplast constriction begins. In a few species like *Cyanidioschyzon merolæ*, chloroplasts have a third plastid-dividing ring located in the chloroplast's intermembrane space.

Late into the constriction phase, dynamin proteins assemble around the outer plastid-dividing ring, helping provide force to squeeze the chloroplast. Meanwhile, the Z-ring and the inner plastid-dividing ring break down. During this stage, the many chloroplast DNA plasmids floating around in the stroma are partitioned and distributed to the two forming daughter chloroplasts.

Later, the dynamins migrate under the outer plastid dividing ring, into direct contact with the chloroplast's outer membrane, to cleave the chloroplast in two daughter chloroplasts.

A remnant of the outer plastid dividing ring remains floating between the two daughter chloroplasts, and a remnant of the dynamin ring remains attached to one of the daughter chloroplasts.

Of the five or six rings involved in chloroplast division, only the outer plastid-dividing ring is present for the entire constriction and division phase—while the Z-ring forms first, constriction does not begin until the outer plastid-dividing ring forms.

Regulation

In species of algae that contain a single chloroplast, regulation of chloroplast division is extremely important to ensure that each daughter cell receives a chloroplast—chloroplasts can't be made from scratch. In organisms like plants, whose cells contain multiple chloroplasts, coordination is looser and less important. It is likely that chloroplast and cell division are somewhat synchronized, though the mechanisms for it are mostly unknown.

Light has been shown to be a requirement for chloroplast division. Chloroplasts can grow and progress through some of the constriction stages under poor quality green light, but are slow to complete division—they require exposure to bright white light to complete division. Spinach leaves grown under green light have been observed to contain many large dumbbell-shaped chloroplasts. Exposure to white light can stimulate these chloroplasts to divide and reduce the population of dumbbell-shaped chloroplasts.

Chloroplast Inheritance

Like mitochondria, chloroplasts are usually inherited from a single parent. Biparental chloroplast inheritance—where plastid genes are inherited from both parent plants—occurs in very low levels in some flowering plants.

Many mechanisms prevent biparental chloroplast DNA inheritance, including selective destruction of chloroplasts or their genes within the gamete or zygote, and chloroplasts from one parent being excluded from the embryo. Parental chloroplasts can be sorted so that only one type is present in each offspring.

Gymnosperms, such as pine trees, mostly pass on chloroplasts paternally, while flowering plants often inherit chloroplasts maternally. Flowering plants were once thought to only inherit chloroplasts maternally. However, there are now many documented cases of angiosperms inheriting chloroplasts paternally.

Angiosperms, which pass on chloroplasts maternally, have many ways to prevent paternal inheritance. Most of them produce sperm cells that do not contain any plastids. There are many other documented mechanisms that prevent paternal inheritance in these flowering plants, such as different rates of chloroplast replication within the embryo.

Among angiosperms, paternal chloroplast inheritance is observed more often in hybrids than in offspring from parents of the same species. This suggests that incompatible hybrid genes might interfere with the mechanisms that prevent paternal inheritance.

Transplastomic Plants

Recently, chloroplasts have caught attention by developers of genetically modified crops. Since, in most flowering plants, chloroplasts are not inherited from the male parent, transgenes in these plastids cannot be disseminated by pollen. This makes plastid transformation a valuable tool for the creation and cultivation of genetically modified plants that are biologically contained, thus posing significantly lower environmental risks. This biological containment strategy is therefore suitable for establishing the coexistence of conventional and organic agriculture. While the reliability of this mechanism has not yet been studied for all relevant crop species, recent results in tobacco plants are promising, showing a failed containment rate of transplastomic plants at 3 in 1,000,000.

Etioplast

Etioplasts are chloroplasts that have not been exposed to light. They are usually found in flowering plants (Angiosperms) grown in the dark. If a plant is kept out of light for several days, its normal chloroplasts will actually convert into etioplasts. Etioplasts lack active pigment and can technically be considered leucoplasts. High concentrations of etioplasts will cause leaves to appear yellow rather than green.

These plant organelles contain prolamellar bodies, which are membrane aggregations of semi-crystalline lattices of branched tubules that carry the precursor pigment for chlorophyll. The prolamellar bodies are often (and presumed always) arranged in geometric patterns.

They are converted to chloroplasts via the stimulation of chlorophyll synthesis by the plant hormone cytokinin soon after exposure to light. Thylakoids and grana arise from the prolamellar bodies during this process.

Gerontoplast

A gerontoplast is a plastid found in formerly green tissues that are currently senescing. A gerontoplast is a chloroplast that has re-purposed through a process of developmental senescence.

Transformation of Chloroplasts to Gerontoplasts

The term *gerontoplast* was first introduced by Sitte (1977) to define the unique features of the plastid formed during leaf senescence. The process of senescence brings about regulated dismantling of cellular organelles. Chloroplast shows the first sign of senescence induced degradation and is the last organelle to survive when other organelles are completely disorganized. The formation of gerontoplast from chloroplast during senescence involves extensive structural modifications of thylakoid membrane with the concomitant formation of a large number of plastoglobuli with lipophilic materials. The envelope of the plastid, however, remains intact.

Chromoplast

The coloration of the petals and sepals on the Bee orchid is controlled by a specialized organelle in plant cells called a chromoplast.

Chromoplasts are plastids, heterogeneousorganelles responsible for pigmentsynthesis and storage in specific photosynthetic eukaryotes. It is thought that like all other plastids including chloroplasts and leucoplasts they are descended from symbioticprokaryotes.

Function

Chromoplasts are found in fruits, flowers, roots, and stressed and aging leaves, and are responsible for their distinctive colors. This is always associated with a massive increase in the accumulation of carotenoid pigments. The conversion of chloroplasts to chromoplasts in ripening is a classic example.

They are generally found in mature tissues and are derived from preexisting mature plastids. Fruits and flowers are the most common structures for the biosynthesis of carotenoids, although other reactions occur there as well including the synthesis of sugars, starches, lipids, aromatic compounds, vitamins and hormones. The DNA in chloroplasts and chromoplasts is identical. One subtle difference in DNA was found after a liquid chromatography analysis of tomato chromoplasts was conducted, revealing increased cytosine methylation.

Chromoplasts synthesize and store pigments such as orange carotene, yellow xanthophylls, and various other red pigments. As such, their color varies depending on what pigment they contain. The main evolutionary purpose of chromoplasts is probably to attract pollinators or eaters of colored fruits, which help disperse seeds. However, they are also found in roots such as carrots and sweet potatoes. They allow the accumulation of large quantities of water-insoluble compounds in otherwise watery parts of plants.

When leaves change color in the autumn, it is due to the loss of green chlorophyll, which unmasks preexisting carotenoids. In this case, relatively little new carotenoid is produced—the change in plastid pigments associated with leaf senescence is somewhat different from the active conversion to chromoplasts observed in fruit and flowers.

There are some species of flowering plants that contain little to no carotenoids. In such cases there are plastids present within the petals that closely resemble chromoplasts and are sometimes visually indistinguishable. Anthocyanins and flavonoids located in the cell vacuoles are responsible for other colors of pigment.

The term "chromoplast" is occasionally used to include *any* plastid that has pigment, mostly to emphasize the difference between them and the various types of leucoplasts, plastids that have no pigments. In this sense, chloroplasts are a specific type of chromoplast. Still, "chromoplast" is more often used to denote plastids with pigments other than chlorophyll.

Structure and Classification

Using a light microscope chromoplasts can be differentiated and are classified into four main types. The first type is composed of proteic stroma with granules. The second is composed of protein crystals and amorphous pigment granules. The third type is composed of protein and pigment crystals. The fourth type is a chromoplast which only contains crystals. An electron microscope reveals even more, allowing for the identification of substructures such as globules, crystals, membranes, fibrils and tubules. The substructures found in chromoplasts are not found in the mature plastid that it divided from.

The presence, frequency and identification of substructures using an electron microscope has led to further classification, dividing chromoplasts into five main categories: Globular chromoplasts, crystalline chromoplasts, fibrillar chromoplasts, tubular chromoplasts and membranous chromoplasts. It has also been found that different types of chromoplasts can coexist in the same organ. Some examples of plants in the various categories include mangos, which have globular chromoplasts, and carrots which have crystalline chromoplasts.

Although some chromoplasts are easily categorized, others have characteristics from multiple categories that make them hard to place. Tomatoes accumulate carotenoids, mainly lycopene crystalloids in membrane-shaped structures, which could place them in either the crystalline or membranous category.

Evolution

Plastids are descendants of cyanobacteria, photosynthetic prokaryotes, which integrated themselves into the eukaryotic ancestor of algæ and plants, forming an endosymbiotic relationship. The ancestors of plastids diversified into a variety of plastid types, including chromoplasts. Plastids also possess their own small genome and some have the ability to produce a percentage of their own proteins.

The main evolutionary purpose of chromoplasts is to attract animals and insects to pollinate their flowers and disperse their seeds. The bright colors often produced by chromoplasts is one of many ways to achieve this. Many plants have evolved symbiotic relationships with a single pollinator. Color can be a very important factor in determining which pollinators visit a flower, as specific colors attract specific pollinators. White flowers tend to attract beetles, bees are most often attracted to violet and blue flowers, and butterflies are often attracted to warmer colors like yellows and oranges.

Research

Chromoplasts are not widely studied and are rarely the main focus of scientific research. They often play a role in research on the tomato plant (*Solanum lycopersicum*). Lycopene is responsible for the red color of a ripe fruit in the cultivated tomato, while the yellow color of the flowers is due to xanthophyllsviolaxanthin and neoxanthin.

Carotenoid biosynthesis occurs in both chromoplasts and chloroplasts. In the chromoplasts of tomato flowers, carotenoid synthesis is regulated by the genes Psyl, Pds, Lcy-b, and Cyc-b. These genes, in addition to others, are responsible for the formation of carotenoids in organs and structures. For example, the Lcy-e gene is highly expressed in leaves, which results in the production of the carotenoid lutein.

White flowers are caused by a recessive allele in tomato plants. They are less desirable in cultivated crops because they have a lower pollination rate. In one study, it was found that chromoplasts are still present in white flowers. The lack of yellow pigment in their petals and anthers is due to a mutation in the CrtR-b2 gene which disrupts the carotenoid biosynthesis pathway.

The entire process of chromoplast formation is not yet completely understood on the molecular level. However, electron microscopy has revealed part of the transformation from chloroplast to chromoplast. The transformation starts with remodeling of the internal membrane system with the lysis of the intergranal thylakoids and the grana. New membrane systems form in organized membrane complexes called thylakoid plexus. The new membranes are the site of the formation of carotenoid crystals. These newly synthesized membranes do not come from the thylakoids, but rather from vesicles generated from the inner membrane of the plastid. The most obvious biochemical change would be the downregulation of photosynthetic gene expression which results in the loss of chlorophyll and stops photosynthetic activity.

In oranges, the synthesis of carotenoids and the disappearance of chlorophyll causes the color of the fruit to change from green to yellow. The orange color is often added artificially—light yellow-orange is the natural color created by the actual chromoplasts.

Valencia oranges *Citris sinensis L* are a cultivated orange grown extensively in the state of Florida. In the winter, Valencia oranges reach their optimum orange-rind color while reverting to a green color in the spring and summer. While it was originally thought that chromoplasts were the final stage of plastid development, in 1966 it was proved that chromoplasts can revert to chloroplasts, which causes the oranges to turn back to green.

Compare

- Plastid

    o Chloroplast and etioplast

    o Chromoplast

    o Leucoplast

        ▪ Amyloplast

        ▪ Elaioplast

        ▪ Proteinoplast

Leucoplast

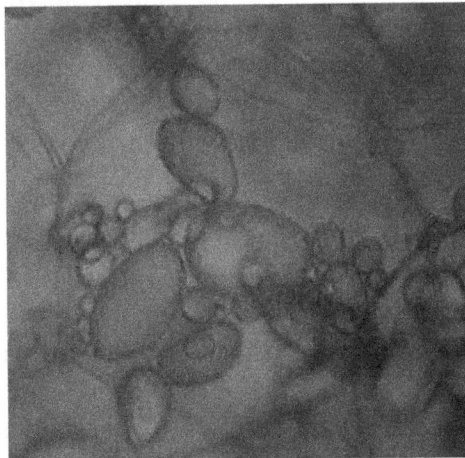

Leucoplasts, specifically, amyloplasts

Leucoplasts are a category of plastid and as such are organelles found in plant cells. They are non-pigmented, in contrast to other plastids such as the chloroplast.

Lacking photosynthetic pigments, leucoplasts are not green and are located in non-photosynthetic tissues of plants, such as roots, bulbs and seeds. They may be specialized for bulk storage of starch, lipid or protein and are then known as amyloplasts, elaioplasts, or proteinoplasts (also called aleuroplasts) respectively. However, in many cell types, leucoplasts do not have a major storage function and are present to provide a wide range of essential biosynthetic functions, including the synthesis of fatty acids such as palmitic acid, many amino acids, and tetrapyrrole compounds such as heme. In general, leucoplasts are much smaller than chloroplasts and have a variable morphology, often described as amoeboid. Extensive networks of stromules interconnecting leucoplasts have been observed in epidermal cells of roots, hypocotyls, and petals, and in callus and suspension culture cells of tobacco. In some cell types at certain stages of development, leucoplasts are clustered around the nucleus with stromules extending to the cell periphery, as observed for proplastids in the root meristem.

Etioplasts, which are pre-granal, immature chloroplasts but can also be chloroplasts that have been deprived of light, lack active pigment and can be considered leucoplasts. After several minutes exposure to light, etioplasts begin to transform into functioning chloroplasts and cease being leucoplasts. Amyloplasts are of large size and store starch. Proteinoplasts store proteins and are found in seeds (pulses). Elaioplasts store fats and oils and are found in seeds. They are also called oleosomes. [castor, groundnut] Etioplasts are plastids without pigments and store food and lamellar structures. These plastids occur in etiolated plants due to the absence of light.

Elaioplast

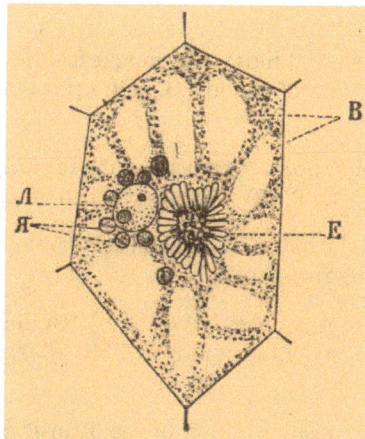

Illustration from Collegiate Dictionary, FA Brockhaus and IA Efron, circa 1905. Cell of very young leaf of *Vanilla planifolia*; E - elaioplasts; Л - the nucleus; Я - leucoplasts; B - vacuoles

Elaioplasts are a type of leucoplast that is specialized for the storage of lipids in plants. Elaioplasts house oil body deposits as rounded plastoglobuli, which are essentially fat droplets.

Being a variety of leucoplast, elaioplasts are non-pigmented and fall into the much broader organelle category of plantplastids. A different example of a storage-specialized leucoplast is the amyloplast, which stores starch.

Proteinoplast

Proteinoplasts (sometimes called *proteoplasts*, *aleuroplasts*, and *aleuronaplasts*) are specialized organelles found only in plant cells. Proteinoplasts belong to a broad category of organelles known as plastids. Because they lack pigment, proteinoplasts are more specifically a kind of leucoplast. They contain crystalline bodies of protein and can be the sites of enzyme activity involving those proteins. Proteinoplasts are found in many seeds, such as brazil nuts, peanuts and pulses. Although all plastids contain high concentrations of protein, proteinoplasts were identified in the 1960s and 1970s as having large protein inclusions that are visible with both light microscopes and electron microscopes.

A book written in 2007 noted that no scientific research had been published in the previous 25 years on proteinoplasts.

References

- Zhang, Q.; Sodmergen (2010). "Why does biparental plastid inheritance revive in angiosperms?". Journal of Plant Research. 123 (2): 201–206. PMID 20052516. doi:10.1007/s10265-009-0291-z

- Jones, Daniel (2003) [1917], Peter Roach, James Hartmann and Jane Setter, eds., English Pronouncing Dictionary, Cambridge: Cambridge University Press, ISBN 3-12-539683-2 CS1 maint: Uses editors parameter (link)

- Viola, R.; Nyvall, P.; Pedersén, M. (2001). "The unique features of starch metabolism in red algae". Proceedings of the Royal society of London, B. 268: 1417–1422. PMC 1088757. PMID 11429143. doi:10.1098/rspb.2001.1644

- Lewis, L. A.; McCourt, R. M. (2004). "Green algae and the origin of land plants". American Journal of Botany. 91 (10): 1535–56. PMID 21652308. doi:10.3732/ajb.91.10.1535

- Burgess, Jeremy (1989). An introduction to plant cell development. Cambridge: Cambridge university press. p. 62. ISBN 0-521-31611-1

- Kumar, K.; Mella-Herrera, R. A.; Golden, J. W. (2010). "Cyanobacterial Heterocysts". Cold Spring Harbor Perspectives in Biology. 2 (4): a000315. PMC 2845205. PMID 20452939. doi:10.1101/cshperspect.a000315

- Soll, J r.; Soll, J (1996). "Phosphorylation of the Transit Sequence of Chloroplast Precursor Proteins". Journal of Biological Chemistry. 271 (11): 6545–54. PMID 8626459. doi:10.1074/jbc.271.11.6545

- Berg JM; Tymoczko JL; Stryer L. (2002). Biochemistry. (5th ed.). W H Freeman. pp. Section 19.4. Retrieved 30 October 2012

- Keeling, Patrick J. (2004). "Diversity and evolutionary history of plastids and their hosts". American Journal of Botany. 91 (10): 1481–93. PMID 21652304. doi:10.3732/ajb.91.10.1481

- J D, Rochaix (1998). The molecular biology of chloroplasts and mitochondria in Chlamydomonas. Dordrecht [u.a.]: Kluwer Acad. Publ. pp. 550–565. ISBN 978-0-7923-5174-0

- Retallack, B; Butler, RD (1970). "The development and structure of pyrenoids in Bulbochaete hiloensis". Journal of Cell Science. 6 (1): 229–41. PMID 5417694

- Schnepf, Eberhard; Elbrächter, Malte (1999). "Dinophyte chloroplasts and phylogeny – A review". Grana. 38 (2–3): 81–97. doi:10.1080/00173139908559217

- Steer, Brian E.S. Gunning, Martin W. (1996). Plant cell biology : structure and function. Boston, Mass.: Jones and Bartlett Publishers. p. 24. ISBN 0-86720-504-0

- Takagi, Shingo (December 2002). "Actin-based photo-orientation movement of chloroplasts in plant cells". Journal of Experimental Biology. 206: 1963–1969. doi:10.1242/jeb.00215. Retrieved 6 March 2013

- Milo, Ron; Philips, Rob. "Cell Biology by the Numbers: How large are chloroplasts?". book.bionumbers.org. Retrieved 7 February 2017

- May, T. (2000). "14-3-3 Proteins Form a Guidance Complex with Chloroplast Precursor Proteins in Plants". The Plant Cell Online. 12: 53–64. doi:10.1105/tpc.12.1.53

- Burgess,, Jeremy (1989). An introduction to plant cell development (Pbk. ed.). Cambridge: Cambridge university press. p. 57. ISBN 0-521-31611-1

5

Essential Aspects of Plant Cell Biology

The essential aspects of plant cell biology are plasmolysis, turgor pressure, oleosin and stoma. Turgor pressure is the force in the cell which helps in pushing the plasma membrane against the cell wall. The aspects elucidated in this chapter are of vital importance, and provide a better understanding of plant cell biology.

Plasmolysis

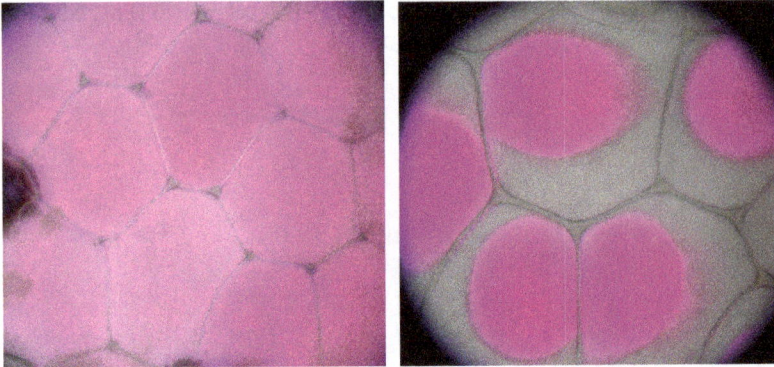

Before plasmolysis (left) and after (right)

Plasmolysis is the process in which cells lose water in a hypertonic solution. The reverse process, cytolysis, can occur if the cell is in a hypotonic solution resulting in a lower external osmotic pressure and a net flow of water into the cell. Through observation of plasmolysis and deplasmolysis, it is possible to determine the tonicity of the cell's environment as well as the rate solute molecules cross the cellular membrane.

Turgidity

A plant cell in hypotonic solution will absorb water by endosmosis, so that the increased volume of water in the cell will increase pressure, making the protoplasm push against the cell wall, a condition known as turgor. Turgor makes plant cells push against each other in the same way and is the main line method of support in non-woody plant tissue. Plant cell walls resist further water entry after a certain point, known as full turgor, which stops plant cells from bursting as animal cells do in the same conditions. This is also the reason that plants stand upright. Without the stiffness of the plant cells the plant would fall under its own weight. Turgor pressure allows plants to stay firm and erect, and plants without turgor pressure (known as flaccid) wilt. A cell begins to decline in turgor pressure only when there is no air spaces surrounding it and eventually leads to a greater osmotic pressure than that of the cell. Vacuoles play a role in turgor pressure when water leaves the cell due to hyperosmotic solutions containing solutes such as mannitol, sorbitol, and sucrose.

Plasmolysis

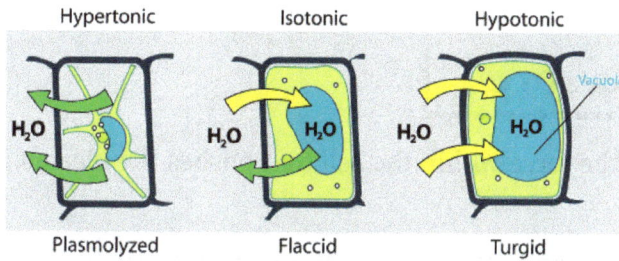

Plant cell under different environments

If a plant cell is placed in a hypertonic solution, the plant cell loses water and hence turgor pressure by plasmolysis: pressure decreases to the point where the protoplasm of the cell peels away from the cell wall, leaving gaps between the cell wall and the membrane and making the plant cell shrink and crumple. A continued decrease in pressure eventually leads to cytorrhysis – the complete collapse of the cell wall. Plants with cells in this condition wilt. After plasmolysis the gap between the cell wall and the cell membrane in a plant cell is filled with hypertonic solution. This is because as the solution surrounding the cell is hypertonic, exosmosis takes place and the space between the cell wall and cytoplasm is filled with solutes, as most of the water drains away and hence the concentration inside the cell becomes more hypertonic. There are some mechanisms in plants to prevent excess water loss in the same way as excess water gain. Plasmolysis can be reversed if the cell is placed in a hypotonic solution. Stomata help keep water in the plant so it does not dry out. Wax also keeps water in the plant. The equivalent process in animal cells is called crenation.

The liquid content of the cell leaks out due to exosmosis. The cell collapses, and the cell membrane pulls away from the cell wall (in plants). Most animal cells consist of only a phospholipid bilayer (plasma membrane) and not a cell wall, therefore shrinking up under such conditions.

Plasmolysis only occurs in extreme conditions and rarely happens in nature. It is induced in the laboratory by immersing cells in strong saline or sugar (sucrose) solutions to cause exosmosis, often using Elodea plants or onion epidermal cells, which have colored cell sap so that the process is clearly visible. Methylene blue can be used to stain plant cells.

Plasmolysis is mainly known as shrinking of cell membrane in hypertonic solution and great pressure.

Plasmolysis can be of two types, either concave plasmolysis or convex plasmolysis. Convex plasmolysis is always irreversible while concave plasmolysis is usually reversible. During concave plasmolysis, the plasma membrane and the enclosed protoplast partially shrinks from the cell wall due to half-spherical, inwarding curving pockets forming between the plasma membrane and the cell wall. During convex plasmolysis, the plasma membrane and the enclosed protoplast shrinks completely from the cell wall, with the plasma membrane's ends in a symmetrically, spherically curved pattern.

Turgor Pressure

Turgor pressure is the force within the cell that pushes the plasma membrane against the cell wall.

It is also called Hydrostatic pressure, and more intricately defined as the pressure measured by a fluid, measured at a certain point within itself when at equilibrium. Generally, turgor pressure is caused by the osmotic flow of water and occurs in plants, fungi, and bacteria. The phenomenon is also observed in protists that have cell walls. This system is not seen in animal cells, seeing how the absence of a cell wall would cause the cell to lyse when under too much pressure. The pressure exerted by the osmotic flow of water is called turgidity. It is caused by the osmotic flow of water through a semipermeable membrane. Osmotic flow of water through a semipermeable membrane is when the water travels from an area with a low-solute concentration, to one with a higher-solute concentration. In plants, this entails the water moving from the low concentration solute outside the cell, into the cell's vacuole.

Mechanism

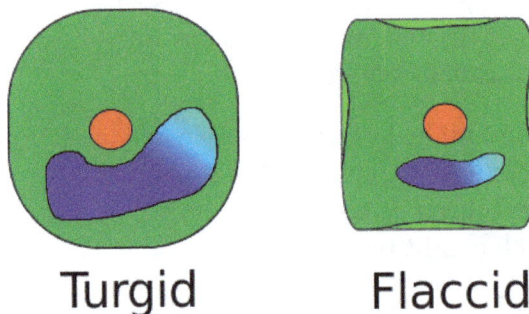

A turgid and flaccid cell

Osmosis is the process in which water flows from an area with a low solute concentration, to an adjacent area with a higher solute concentration until equilibrium between the two areas is reached. All cells are surrounded by a lipid bi-layer cell membrane which permits the flow of water in and out of the cell while also limiting the flow of solutes. In hypertonic solutions, water flows out of the cell which decreases the cell's volume. When in a hypotonic solution, water flows into the membrane and increases the cell's volume. When in an isotonic solution, water flows in and out of the cell at an equal rate.

Turgidity is the point at which the cell's membrane pushes against the cell wall, which is when turgor pressure is high. When the cell membrane has low turgor pressure then it is flaccid. In plants, this is shown as wilted anatomical structures. This is more specifically known as plasmolysis.

The volume and geometry of the cell affects the value of turgor pressure, and how it can have an effect on the cell wall's plasticity. Studies have shown how smaller cells experience a stronger elastic change when compared to larger cells.

Turgor pressure also plays a key role in plant cell growth where the cell wall undergoes irreversible expansion due to the force of turgor pressure as well as structural changes in the cell wall that alter its extensibility.

Turgor Pressure in Plants

Turgor pressure within cells is regulated by osmosis and also causes the cell wall to expand during growth. Along with size, rigidity of the cell is also caused by turgor pressure; a lower pressure results in a wilted cell or plant structure (i.e. leaf, stalk). One mechanism in plants that regulate turgor pressure is its semipermeable membrane, which only allows some solutes to travel in and out of the cell, which can also maintain a minimum amount of pressure. Other mechanisms include transpiration, which results in water loss and decreases turgidity in cells. Turgor pressure is also a large factor for nutrient transport throughout the plant. Cells of the same organism can have differing turgor pressures throughout the organism's structure. In higher plants, turgor pressure is responsible for apical growth of things such as root tips and pollen tubes.

Dispersal

Transport proteins that pump solutes into the cell can be regulated by cell turgor pressure. Lower values allow for an increase in the pumping of solutes; which in turn increases osmotic pressure. This function is important as a plant response when under drought conditions (seeing as turgor pressure is maintained), and for cells which need to accumulate solutes (i.e. developing fruits).

Flowering and Reproductive Organs

It has been recorded that the petals of *Gentiana kochiana* and *Kalanchoe blossfeldiana* bloom via volatile turgor pressure of cells on the plant's adaxial surface. During processes like antherdehiscence, it has been observed that drying endothecium cells cause an outward bending force which led to the release of pollen. This means that lower turgor pressures are observed in these structures due to the fact that they are dehydrated.

Pollen tubes are cells which elongate when pollen lands on the stigma, at the carpal tip. These cells grow rather quickly due to increases turgor pressure. These cells undergo tip growth. The pollen tube of Lilies can have a turgor pressure of 0-21 MPa when growing during this process.

Mature Squirting Cucumber Fruit

Seed Dispersal

In fruits such as *Impatiens parviflora, Oxalia acetosella* and *Ecballium elaterium,* turgor pressure is the method by which seeds are dispersed. In *Ecballium elaterium,* or squirting cucumber, turgor pressure builds up in the fruit to the point that aggressively detaches from the stalk, and seeds and water are squirted everywhere as the fruit falls to the ground. Turgor pressure within the fruit ranges from .003-1.0 MPa.

Growth

Tree Roots Penetrating Rock

Turgor pressure's actions on extensible cell walls is usually said to be the driving force of growth within the cell. An increase of turgor pressure causes expansion of cells and extension of apical cells, pollen tubes, and in other plant structures such as root tips. Cell expansion and an increase in turgor pressure is due to inward diffusion of water into the cell, and turgor pressure increases due to the increasing volume of vacuolarsap. A growing root cell's turgor pressure can be up to 0.6 MPa, which is over three times that of a car tire. Epidermal cells in a leaf can have pressures ranging from 1.5-2.0 MPa. Seeing that plants can operate at such high pressures, it can explain why trees can grow through asphalt and other hard surfaces.

Turgidity

This is observed in a cell where the cell membrane is pushed against the cell wall. In some plants, their cell walls loosen at a quicker rate than water can cross the membrane, which results in a cell with lower turgor pressure.

Stomata

Open Stomata on the Left and Closed Stomata on the Right

Turgor pressure within the stomata regulates when the stomata can open and close, which has a play in transpiration rates of the plant. This is also important because this function regulates water loss within the plant. Lower turgor pressure can mean that the cell has a low water concentration and closing the stomata would help to preserve water. High turgor pressure keeps the stomata open for gas exchanges necessary for photosynthesis.

Mimosa Pudica

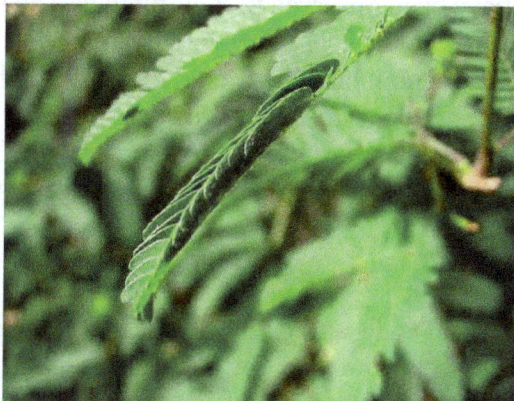

Mimosa pudica

It has been concluded that loss of turgor pressure within the leaves of *Mimosa pudica* is responsible for the reaction the plant has when touched. Other factors such as changes in osmotic pressure, protoplasmic contraction and increase in cellular permeability have been observed to affect this response. It has also been recorded that turgor pressure is different in the upper and lower pulvinar cells of the plant, and the movement of potassium and calcium ions throughout the cells cause the increase in turgor pressure. When touched, the pulvinus is activated and exudes contractile proteins, which in turn increases turgor pressure and closes the leaves of the plant.

Function in Other Taxa

As earlier stated, turgor pressure can be found in other organisms besides plants, and can play a large role in the development, movement and nature of said organisms.

Fungi

Shaggy Ink Caps Bursting through Asphalt due to High Turgor Pressures

In fungi, tugor pressure has been observed as a large factor in substrate penetration. In species such as *Saprolegnia ferax, Magnaporthe grisea* and *Aspergillus oryzae,* immense turgor pressures have been observed in their hyphae. The study showed that they could penetrate substances like plant cells, and synthetic materials such as polyvinyl chloride. In observations of this phenomenon, it is noted that invasive hyphal growth is due to turgor pressure, along with the coenzymes secreted by the fungi to invade said substrates. Hyphal growth is directed related to turgor pressure, and growth slows as turgor pressure decreases. In *Magnaporthe grisea*, pressures of up to 8 MPa have been observed.

Protists

Some protists do not have cell walls and cannot experience turgor pressure. These few protists are ones that use their contractile vacuole to regulate the quantity of water within the cell. Protist cells avoid lysing in solutions by utilizing a vacuole which pumps water out of the cells to maintain osmotic equilibrium.

Animals

Turgor pressure is not observed in animalcells because they lack a cell wall. In organisms with cell walls, the cell wall prevents the cell from lysing from high pressure values.

Diatoms

In Diatoms, the Heterokontophyta have polyphyletic turgor-resistant cell walls. Throughout these organisms' life cycle, carefully controlled turgor pressure is responsible for cell expansion and for the release of sperm, but not for things such as seta growth.

Cyanobacteria

Gas-vaculate cyanobacterium are the ones generally responsible for water-blooms. They have the ability to float due to the accumulation of gases within their vacuole, and the role of turgor pressure and its effect on the capacity of these vacuoles has been observed in varying scientific papers. It is noted that the higher the turgor pressure, the lower the capacity of the gas-vacuoles in different cyanobacterium. Experiments used to correlate osmosis and turgor pressure in prokaryotes have been used to show how diffusion of solutes into the cell have a play on turgor pressure within the cell.

Measurements

When measuring turgor pressure in plants, many things have to be taken into account. It is generally stated that fully turgid cells have a turgor pressure value which is equal to that of the cell, and that flaccid cells have a value at or near zero. Other cellular mechanisms taken into consideration include the protoplast, solutes within the protoplast (solute potential), transpiration rates of the plant, and the tension of cell walls. Measurement is limited depending on the method used, some of which are explored and explained below. Not all methods can be used for all organisms, due to size and other properties. For example, a diatom won't have the same properties as a plant, which would place constrictions on what could be used to infer turgor pressure.

Units

Units used to measure turgor pressure are independent from the measures used to infer its values. Common units include bars, MPa, or newtons. 1 bar is equal to .1 MPa.

Methods

Water Potential Equation

Turgor pressure can be deduced when total water potential, Ψ_w, and osmotic potential, Ψ_s, are known in a water potential equation. These equations are used to measure the total water potential of a plant by using variables such as matric potential, osmotic potential, pressure potential, gravitational effects and turgor pressure. After taking the difference between Ψ_s and Ψ_w, the value for turgor pressure is given. When using this method, gravity and matric potential are considered to be negligible, since their values are generally either negative or close to zero.

Pressure-Bomb Technique

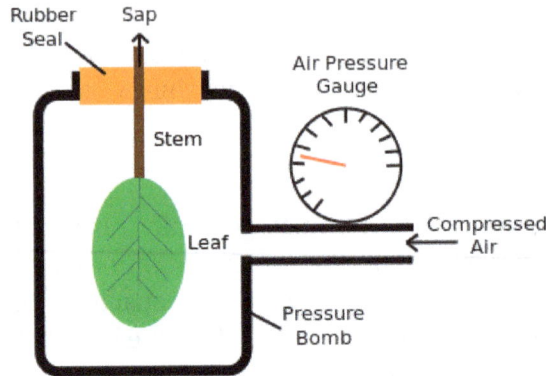

Diagram of a Pressure Bomb

The pressure bomb was developed by plant physiologist Lan Wang and colleagues in order to test water movement through plants. The instrument is used to measure turgor pressure by placing a leaf (with stem attached) into a closed chamber where pressurized gas is added in increments. Measurements are taken when xylem sap appears out of the cut surface and at the point which it doesn't accumulae or retreat back into the cut surface.

Atomic Force Microscope

Atomic Force Microscopes use a type of Scanning probe microscopy (SPM). Small probes are introduced to the area of interest, and a spring within the probe measures values via displacement. This method can be used to measure turgor pressure of organisms. When using this method, supplemental information such as continuum mechanic equations, single force depth curves and cell geometries can be used to quantify turgor pressures within a given area (usually a cell).

Pressure Probe

This machine was originally used to measure individual algal cells, but can now be used on larger-celled specimens. It is usually used on higher plant tissues, but wasn't used to measure turgor pressure until Hüsken and Zimmerman improved on the method. Pressure probes measure turgor pressure via displacement. A glass micro-capillary tube is inserted into the cell and whatever the cell exudes into the tube is observed through a microscope. An attached device then measures how much pressure is required to push the emission back into the cell.

Micro-manipulation Probe

These are used to accurately quantify measurements of smaller cells. In an experiment by Weber, Smith and colleagues, single tomato cells were compressed between a micro-manipulation probe and glass to allow the pressure probe's micro-capillary to find the cell's turgor pressure.

Theoretical Speculations of Turgor Pressure

'Negative' Turgor Pressure

It has been observed that the value of Ψ_w decreases as the cell becomes more dehydrated, but scientists have speculated whether this value will continue to decrease but never fall to zero, or if the value can be less than zero. There have been studies which show that negative cell pressures can exist in xerophytic plants, but a paper by M. T. Tyree explores the notion of if this is actually possible, or just a conclusion based on misinterpreted data. In his paper, he concludes that by miscategorizing "bound" and "free" water in a cell, researchers that claimed to have found negative turgor pressure values were incorrect. By analyzing isotherms of apoplastic and symplastic water, he shows that negative turgor pressures cannot be present within arid plants due to net water loss of the specimen during droughts. Of course, his is just this analyzation and interpretation of data, and negative turgor pressure values are still used within the scientific community.

Tip Growth in Higher Plants

An hypothesis formed by M. Harold and his colleagues suggests that tip growth in higher plans is amoebic in nature, and isn't caused by turgor pressure as is widely believed, meaning that extension is caused by the actin cytoskeleton in these plant cells. Regulation of cell growth is implied to be caused by cytoplasmic micro-tubules which control the orientation of cellulose fibrils, which are deposited into the adjacent cell wall and results in growth. In plants, the cells are surrounded by cell walls and filamentous proteins which retain and adjust the plant cell's growth and shape. As explained in the paper, lower plants grow through apical growth, which differs since the cell wall only expands on one end of the cell.

Oleosin

Oleosins are structural proteins found in vascular plant oil bodies and in plant cells. Oil bodies are not considered organelles because they have a single layer membrane and lack the pre-requisite double layer membrane in order to be considered an organelle. They are found in plant parts with high oil content that undergo extreme desiccation as part of their maturation process, and help stabilize the bodies.

Oleosins are proteins of 16 kDa to 24 kDa and are composed of three domains: an N-terminalhydrophilic region of variable length (from 30 to 60 residues); a central hydrophobicdomain of about 70 residues and a C-terminal amphipathic region of variable length (from 60 to 100 residues). The central hydrophobic domain is proposed to be made up of beta-strand structure and to interact with the lipids. It is the only domain whose sequence is conserved. Models show oleosins having a hairpin-like hydrophobic shape that is inserted inside the triacylglyceride (TAG), while the hydrophilic parts are left outside oil bodies.

Oleosins have been found on oil bodies of seeds, tapetum cells, and pollen but not fruits. Instead of a stabilizer of oil bodies, oleosins are believed to be involved in water-uptaking of pollen on stigma.

Use in Purification of Recombinant Protein

Oleosins provide an easy way of purifying proteins which have been produced recombinantly in plants. If the protein is made as a fusion protein with oleosin and a protease recognition site is incorporated between them, the fusion protein will sit in the membrane of the oil body, which can be easily isolated by centrifugation. The oil droplets can then be mixed with aqueous medium again, and oleosin cleaved from the protein of interest. Centrifugation will cause two phases to separate again, and the aqueous medium now contains the purified protein.

Stoma

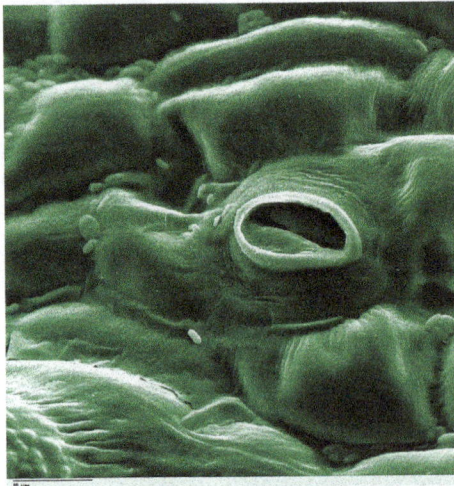

Stoma in a tomato leaf shown via colorized scanning electron microscope image

In botany, a stoma (plural "stomata"), also called a stomate (plural "stomates") (from Greek "mouth"), is a pore, found in the epidermis of leaves, stems, and other organs, that facilitates gas exchange. The pore is bordered by a pair of specialized parenchyma cells known as guard cells that are responsible for regulating the size of the stomatal opening.

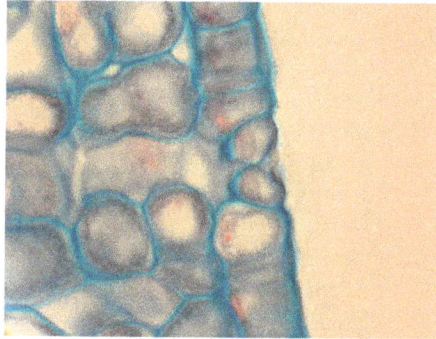

A stoma in cross section

The term is usually used collectively to refer to the entire stomatal complex, consisting of the paired guard cells and the pore itself, which is referred to as the stomatal aperture. Air enters the plant through these openings by gaseous diffusion, and contains carbon dioxide and oxygen, which are used in photosynthesis and respiration, respectively. Oxygen produced as a by-product of photosynthesis diffuses out to the atmosphere through these same openings. Also, water vapor diffuses through the stomata into the atmosphere in a process called transpiration.

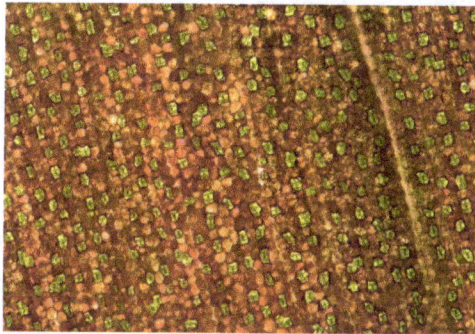

The underside of a leaf. In this species (*Tradescantia zebrina*) the guard cells of the stomata are green because they contain chlorophyll while the epidermal cells are chlorophyll-free and contain red pigments.

Stomata are present in the sporophyte generation of all land plant groups except liverworts. In vascular plants the number, size and distribution of stomata varies widely. Dicotyledons usually have more stomata on the lower surface of the leaves than the upper surface. Monocotyledons such as onion, oat and maize may have about the same number of stomata on both leaf surfaces. In plants with floating leaves, stomata may be found only on the upper epidermis and submerged leaves may lack stomata entirely. Most tree species have stomata only on the lower leaf surface. Leaves with stomata on both the upper and lower leaf are called amphistomatous leaves; leaves with stomata only on the lower surface are hypostomatous, and leaves with stomata only on the upper surface are epistomatous or hyperstomatous. Size varies across species, with end-to-end lengths ranging from 10 to 80 μm and width ranging from a few to 50 μm.

Function

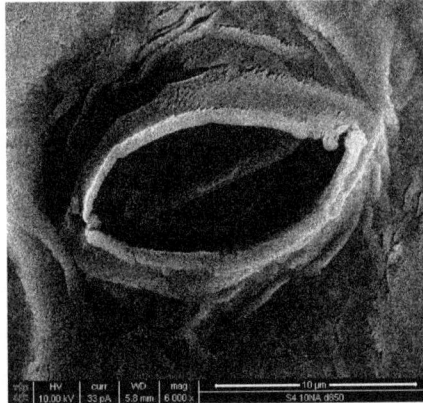

Electron micrograph of a stoma from a *Brassica chinensis* (Bok Choy) leaf

CO_2 Gain and Water Loss

Carbon dioxide, a key reactant in photosynthesis, is present in the atmosphere at a concentration of about 400 ppm. Most plants require the stomata to be open during daytime. The air spaces in the leaf are saturated with water vapour, which exits the leaf through the stomata; this is known as transpiration. Therefore, plants cannot gain carbon dioxide without simultaneously losing water vapour.

Alternative Approaches

Ordinarily, carbon dioxide is fixed to ribulose-1,5-bisphosphate (RuBP) by the enzyme RuBisCO in mesophyll cells exposed directly to the air spaces inside the leaf. This exacerbates the transpiration problem for two reasons: first, RuBisCo has a relatively low affinity for carbon dioxide, and second, it fixes oxygen to RuBP, wasting energy and carbon in a process called photorespiration. For both of these reasons, RuBisCo needs high carbon dioxide concentrations, which means wide stomatal apertures and, as a consequence, high water loss.

Narrower stomatal apertures can be used in conjunction with an intermediary molecule with a high carbon dioxide affinity, PEPcase (Phosphoenolpyruvate carboxylase). Retrieving the products of carbon fixation from PEPCase is in an energy-intensive process, however. As a result, the PEPCase alternative is preferable only where water is limiting but light is plentiful, or where high temperatures increase the solubility of oxygen relative to that of carbon dioxide, magnifying RuBisCo's oxygenation problem.

CAM Plants

A group of mostly desert plants called "CAM" plants (Crassulacean acid metabolism, after the family Crassulaceae, which includes the species in which the CAM process was first discovered) open their stomata at night (when water evaporates more slowly from leaves for a given degree of stomatal opening), use PEPcarboxylase to fix carbon dioxide and store the products in large vacuoles. The following day, they close their stomata and release the carbon dioxide fixed the previous night into the presence of RuBisCO. This saturates RuBisCO with carbon dioxide, allowing minimal photorespiration. This approach, however, is severely limited by the capacity to store fixed carbon in the vacuoles, so it is preferable only when water is severely limiting.

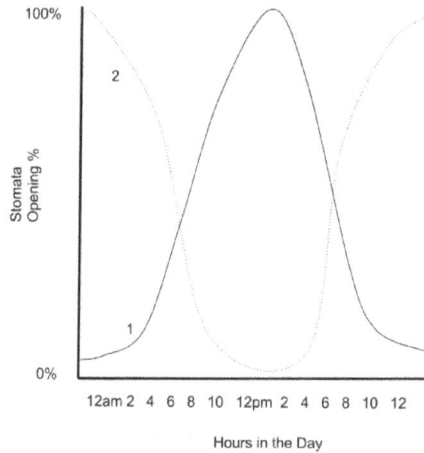

C3 and C4 plants(1) stomata stay open all day and close at night. CAM plants(2) stomata open during the morning and close slightly at noon and then open again in the morning.

Opening and Closure

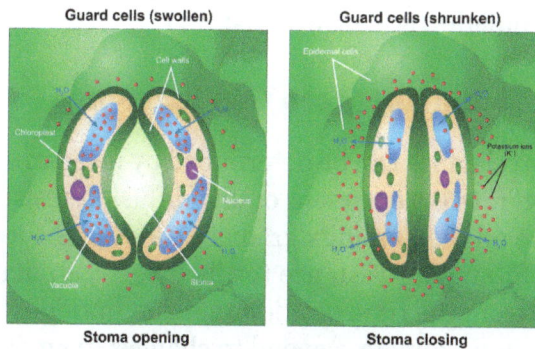

Opening and closing of stoma.

However, most plants do not have the aforementioned facility and must therefore open and close their stomata during the daytime, in response to changing conditions, such as light intensity, humidity, and carbon dioxide concentration. It is not entirely certain how these responses work. However, the basic mechanism involves regulation of osmotic pressure.

When conditions are conducive to stomatal opening (e.g., high light intensity and high humidity), a proton pump drives protons (H^+) from the guard cells. This means that the cells' electrical potential becomes increasingly negative. The negative potential opens potassium voltage-gated channels and so an uptake of potassium ions (K^+) occurs. To maintain this internal negative voltage so that entry of potassium ions does not stop, negative ions balance the influx of potassium. In some cases, chloride ions enter, while in other plants the organic ion malate is produced in guard cells. This increase in solute concentration lowers the water potential inside the cell, which results in the diffusion of water into the cell through osmosis. This increases the cell's volume and turgor pressure. Then, because of rings of cellulose microfibrils that prevent the width of the guard cells from swelling, and thus only allow the extra turgor pressure to elongate the guard cells, whose ends are held firmly in place by surrounding epidermal cells, the two guard cells lengthen by bowing apart from one another, creating an open pore through which gas can move.

When the roots begin to sense a water shortage in the soil, abscisic acid (ABA) is released. ABA binds to receptor proteins in the guard cells' plasma membrane and cytosol, which first raises the pH of the cytosol of the cells and cause the concentration of free Ca^{2+} to increase in the cytosol due to influx from outside the cell and release of Ca^{2+} from internal stores such as the endoplasmic reticulum and vacuoles. This causes the chloride (Cl^-) and inorganic ions to exit the cells. Second, this stops the uptake of any further K^+ into the cells and, subsequently, the loss of K^+. The loss of these solutes causes an increase in water potential, which results in the diffusion of water back out of the cell by osmosis. This makes the cell plasmolysed, which results in the closing of the stomatal pores.

Guard cells have more chloroplasts than the other epidermal cells from which guard cells are derived. Their function is controversial.

Inferring Stomatal Behavior from Gas Exchange

The degree of stomatal resistance can be determined by measuring leaf gas exchange of a leaf. The transpiration rate is dependent on the diffusion resistance provided by the stomatal pores, and also on the humidity gradient between the leaf's internal air spaces and the outside air. Stomatal resistance (or its inverse, stomatal conductance) can therefore be calculated from the transpiration rate and humidity gradient. This allows scientists to investigate how stomata respond to changes in environmental conditions, such as light intensity and concentrations of gases such as water vapor, carbon dioxide, and ozone. Evaporation (E) can be calculated as;

$$E = (e_i - e_a) / Pr$$

where e_i and e_a are the partial pressures of water in the leaf and in the ambient air, respectively, P is atmospheric pressure, and r is stomatal resistance. The inverse of r is conductance to water vapor (g), so the equation can be rearranged to;

$$E = (e_i - e_a)g / P$$

and solved for g;

$$g = EP / (e_i - e_a)$$

Photosynthetic CO_2 assimilation (A) can be calculated from

$$A = (C_a - C_i)g / 1.6P$$

where C_a and C_i are the atmospheric and sub-stomatal partial pressures of CO_2, respectively. The rate of evaporation from a leaf can be determined using a photosynthesis system. These scientific instruments measure the amount of water vapour leaving the leaf and the vapor pressure of the ambient air. Photosynthetic systems may calculate water use efficiency (A/E), g, intrinsic water use efficiency (A/g), and C_i. These scientific instruments are commonly used by plant physiologists to measure CO_2 uptake and thus measure photosynthetic rate.

Evolution

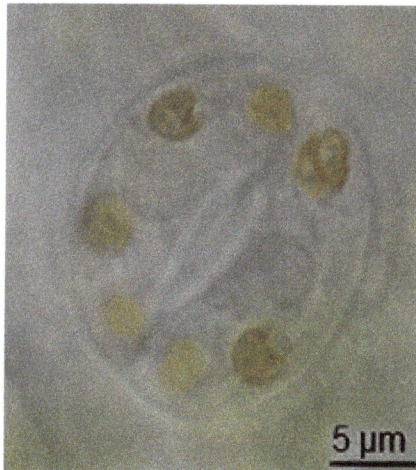

Tomato stoma observed through immersion oil

There is little evidence of the evolution of stomata in the fossil record, but they had appeared in land plants by the middle of the Silurian period. They may have evolved by the modification of conceptacles from plants' alga-like ancestors. However, the evolution of stomata must have happened at the same time as the waxy cuticle was evolving – these two traits together constituted a major advantage for early terrestrial plants.

Development

There are three major epidermal cell types which all ultimately derive from the outermost (L1) tissue layer of the shoot apical meristem, called protodermal cells: trichomes, pavement cells and guard cells, all of which are arranged in a non-random fashion.

An asymmetrical cell division occurs in protodermal cells resulting in one large cell that is fated to become a pavement cell and a smaller cell called a meristemoid that will eventually differentiate into the guard cells that surround a stoma. This meristemoid then divides asymmetrically one to three times before differentiating into a guard mother cell. The guard mother cell then makes one symmetrical division, which forms a pair of guard cells. Cell division is inhibited in some cells so there is always at least one cell between stomata.

Stomatal patterning is controlled by the interaction of many signal transduction components such as *EPF* (Epidermal Patterning Factor), *ERL* (ERecta Like) and *YODA* (a putative MAP kinase kinase kinase). Mutations in any one of the genes which encode these factors may alter the development of stomata in the epidermis. For example, a mutation in one gene causes more stomata that are clustered together, hence is called Too Many Mouths (*TMM*).Whereas, disruption of the *SPCH* (SPeecCHless) gene prevents stomatal development all together. Activation of stomatal production can occur by the activation of EPF1, which activates TMM/ERL, which together activate YODA. YODA inhibits SPCH, causing SPCH activity to decrease, allowing for asymmetrical cell division that initiates stomata formation. Stomatal development is also coordinated by the cellular peptide signal called stomagen, which signals the inhibition of the SPCH, resulting in increased number of stomata.

Environmental and hormonal factors can affect stomatal development. Light increases stomatal development in plants; while, plants grown in the dark have a lower amount of stomata. Auxin represses stomatal development by affecting their development at the receptor level like the ERL and TMM receptors. However, a low concentration of auxin allows for equal division of a guard mother cell and increases the chance of producing guard cells.

Stomatal Crypts

Stomatal crypts are sunken areas of the leaf epidermis which form a chamber-like structure that contains one or more stomata and sometimes trichomes or accumulations of wax. Stomatal crypts can be an adaption to drought and dry climate conditions when the stomatal crypts are very pronounced. However, dry climates are not the only places where they can be found. The following plants are examples of species with stomatal crypts or antechambers: *Nerium oleander*, conifers, and *Drimys winteri* which is a species of plant found in the cloud forest.

Stomata as Pathogenic Pathways

Stomata are obvious holes in the leaf by which, as was presumed for a while, pathogens can enter unchallenged. However, it has been recently shown that stomata do in fact sense the presence of some, if not all, pathogens. However, with the virulent bacteria applied to *Arabidopsis* plant leaves in the experiment, the bacteria released the chemical coronatine, which forced the stomata open again within a few hours.

Stomata and Climate Change

Response of Stomata to Environmental Factors

Photosynthesis, plant water transport (xylem) and gas exchange are regulated by stomatal function which is important in the functioning of plants. Stomatal density and aperture (length of stomata) varies under a number of environmental factors such as atmospheric CO_2 concentration, light intensity, air temperature and photoperiod (daytime duration).

Decreasing stomatal density is one way plants have responded to the increase in concentration of atmospheric CO_2 ($[CO_2]_{atm}$). Although changes in $[CO_2]_{atm}$ response is the least understood mechanistically, this stomatal response has begun to plateau where it is soon expected to impact transpiration and photosynthesis processes in plants.

Future Adaptations during Climate Change

It is expected for $[CO_2]_{atm}$ to reach 500–1000 ppm by 2100. 96% of the past 400 000 years experienced below 280 ppm CO_2 levels. From this figure, it is highly probable that genotypes of today's plants diverged from their pre-industrial relative.

The gene *HIC* (high carbon dioxide) encodes a negative regulator for the development of stomata in plants. Research into the *HIC* gene using *Arabidopsis thaliana* found no increase of stomatal development in the dominant allele, but in the 'wild type' recessive allele showed a large increase, both in response to rising CO_2 levels in the atmosphere. These studies imply the plants response to changing CO_2 levels is largely controlled by genetics.

Agricultural Implications

In the face of ecological contingencies such as increasing temperatures, changes in rainfall patterns, long term climate change, and biotic influences of human management interventions, it is expected to reduce the production and quality of food and have a negative impact on agricultural production.

The CO_2 fertiliser effect has been greatly overestimated during Free-Air Carbon dioxide Enrichment (FACE) experiments where results show increased CO_2 levels in the atmosphere enhances photosynthesis, reduce transpiration, and increase water use efficiency (WUE). Increased biomass is one of the effects with simulations from experiments predicting a 5–20% increase in crop yields at 550 ppm of CO_2. Rates of leaf photosynthesis were shown to increase by 30–50% in C3 plants, and 10–25% in C4 under doubled CO_2 levels. The existence of a feedback mechanism results a phenotypic plasticity in response to $[CO_2]_{atm}$ that may have been an adaptive trait in the evolution of plant respiration and function.

Predicting how stomata perform during adaptation is useful for understanding the productivity of plant systems for both natural and agricultural systems. Plant breeders and farmers are beginning to work together using evolutionary and participatory plant breeding to find the best suited species such as heat and drought resistant crop varieties that could naturally evolve to the change in the face of food security challenges.

Cytokinesis

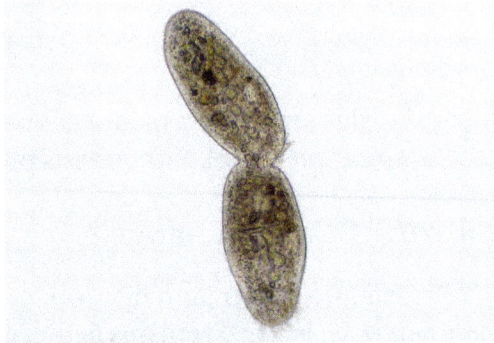

Cilliate undergoing cytokinesis, with the cleavage furrow being clearly visible

Cytokinesis is that part of the cell division process during which the cytoplasm of a single eukaryotic cell divides into two daughter cells. Cytoplasmic division begins during or after the late stages of nuclear division in mitosis and meiosis. During cytokinesis the spindle apparatus partitions and transports duplicated chromatids into the cytoplasm of the separating daughter cells. It thereby ensures that chromosome number and complement are maintained from one generation to the next and that except in special cases the daughter cells will be functional copies of the parent cell. After the completion of the telophase and cytokinesis each daughter cell enters the interphase of the cell cycle.

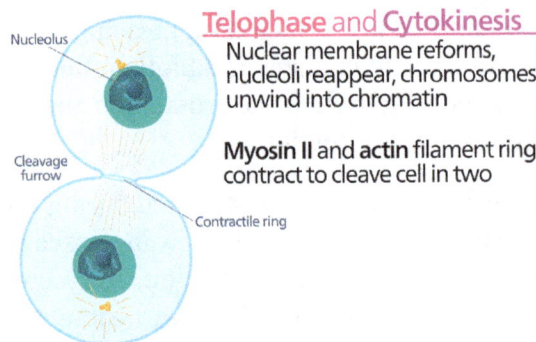

Animal cell telophase and cytokinesis

Particular functions demand various deviations from the process of symmetrical cytokinesis; for example in oogenesis in animals the ovum takes almost all the cytoplasm and organelles. This leaves very little for the resulting polar bodies, which in most species die without function, though they do take on various special functions in other species. Another form of mitosis occurs in tissues such as liver and skeletal muscle; it omits cytokinesis, thereby yielding multinucleate cells.

Plant cytokinesis differs from animal cytokinesis, partly because of the rigidity of plant cell walls. Instead of plant cells forming a cleavage furrow such as develops between animal daughter cells, a dividing structure known as the cell plate forms in the cytoplasm and grows into a new, doubled cell wall between plant daughter cells.

Cytokinesis largely resembles the prokaryotic process of binary fission, but because of differences between prokaryotic and eukaryotic cell structures and functions, the mechanisms differ. For instance a bacterial cell has only a single chromosome in the form of a closed loop, in contrast to the linear, usually multiple, chromosomes of eukaryotes; accordingly bacteria construct no mitotic spindle in cell division. Also, duplication of prokaryotic DNA takes place during the actual separation of chromosomes; in mitosis duplication takes place during the interphase before mitosis begins, though the daughter chromatids do not separate completely before the anaphase.

Plant Cell Cytokinesis

Due to the presence of a cell wall, cytokinesis in plant cells is significantly different from that in animal cells, Rather than forming a contractile ring, plant cells construct a cell plate in the middle of the cell. The stages of cell plate formation include (1) creation of the phragmoplast, an array of microtubules that guides and supports the formation of the cell plate; (2) trafficking of vesicles to the division plane and their fusion to generate a tubular-vesicular network; (3) continued fusion of membrane tubules and their transformation into membrane sheets upon the deposition of callose, followed by deposition of cellulose and other cell wall components; (4) recycling of excess membrane and other material from the cell plate; and (5) fusion with the parental cell wall.

The phragmoplast is assembled from the remnants of the mitotic spindle, and serves as a track for the trafficking of vesicles to the phragmoplast midzone. These vesicles contain lipids, proteins and carbohydrates needed for the formation of a new cell boundary. Electron tomographic studies have identified the Golgi apparatus as the source of these vesicles, but other studies have suggested that they contain endocytosed material as well.

These tubules then widen and fuse laterally with each other, eventually forming a planar, fenestrated sheet . As the cell plate matures, large amounts of membrane material are removed via clathrin-mediated endocytosis Eventually, the edges of the cell plate fuse with the parental plasma membrane, often in an asymmetrical fashion, thus completing cytokinesis. The remaining fenestrae contain strands of endoplasmic reticulum passing through them, and are thought to be the precursors of plasmodesmata.

The construction of the new cell wall begins within the lumen of the narrow tubules of the young cell plate. The order in which different cell wall components are deposited has been determined largely by immuno-electron microscopy. The first components to arrive are pectins, hemicelluloses, and arabinogalactan proteins carried by the secretory vesicles that fuse to form the cell plate. The next component to be added is callose, which is polymerized directly at the cell plate by callose synthases. As the cell plate continues to mature and fuses with the parental plasma membrane, the callose is slowly replaced with cellulose, the primary component of a mature cell wall. The middle lamella (a glue-like layer containing pectin) develops from the cell plate, serving to bind the cell walls of adjoining cells together.

Forces in Cytokinesis

Animal Cells

Cytokinetic furrow ingression is powered by Type II Myosin ATPase. Since Myosins are recruited to the medial region, the contractile forces acting on the cortex resemble a 'purse string' constriction pulling inwards. This leads to the inward constriction. The plasma membrane by virtue of its close association with the cortex via crosslinker proteins.

Cell Plate

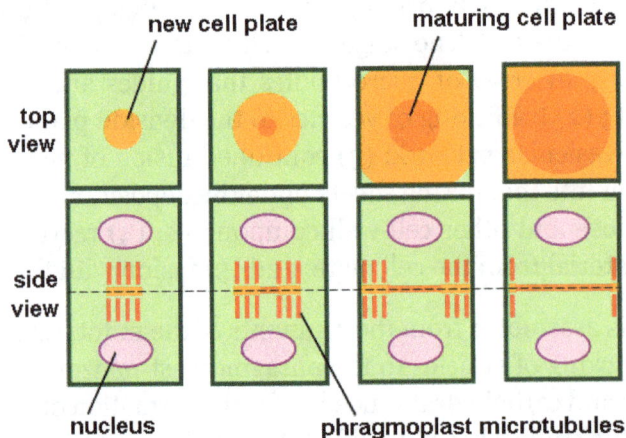

Phragmoplast and cell plate formation in a plant cell during cytokinesis. Left side: Phragmoplast forms and cell plate starts to assemble in the center of the cell. Towards the right: Phragmoplast enlarges in a donut-shape towards the outside of the cell, leaving behind mature cell plate in the center. The cell plate will transform into the new cell wall once cytokinesis is complete.

Cytokinesis in terrestrial plants occurs by cell plate formation. This process entails the delivery of Golgi-derived and endosomal vesicles carrying cell wall and cell membrane components to the plane of cell division and the subsequent fusion of these vesicles within this plate.

After formation of an early tubulo-vesicular network at the center of the cell, the initially labile cell plate consolidates into a tubular network and eventually a fenestrated sheet. The cell plate grows outward from the center of the cell to the parental plasma membrane with which it will fuse, thus completing cell division. Formation and growth of the cell plate is dependent upon the phragmoplast, which is required for proper targeting of Golgi-derived vesicles to the cell plate.

As the cell plate matures in the central part of the cell, the phragmoplast disassembles in this region and new elements are added on its outside. This process leads to a steady expansion of the phragmoplast and, concomitantly, to a continuous retargeting of Golgi-derived vesicles to the growing edge of the cell plate. Once the cell plate reaches and fuses with the plasma membrane the phragmoplast disappears. This event not only marks the separation of the two daughter cells, but also initiates a range of biochemical modifications that transform the callose-rich, flexible cell plate into a cellulose-rich, stiff primary cell wall.

The heavy dependence of cell plate formation on active Golgi stacks explains why plant cells, unlike mammalian cells, do not disassemble their secretion machinery during cell division.

Phragmosome

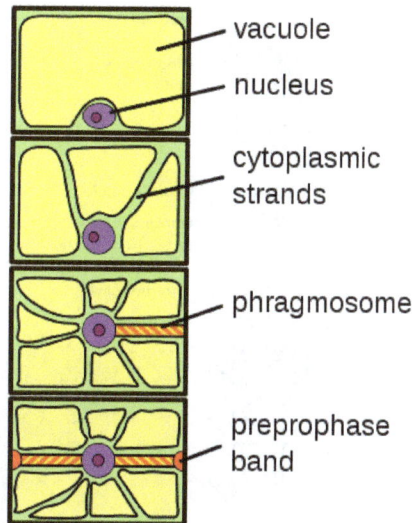

Phragmosome formation in a highly vacuolated plant cell. From top to bottom: 1) Interphase cell with large central vacuole. 2) Cytoplasmic strands starting to penetrate vacuole. 3) Nucleus migration into center and formation of the phragmosome. 4) Phragmosome formation completed and formation of preprophase band marking future cell division plane.

The phragmosome is a sheet of cytoplasm forming in highly vacuolated plant cells in preparation for mitosis. In contrast to animal cells, plant cells often contain large central vacuoles occupying up to 90% of the total cell volume and pushing the nucleus against the cell wall. In order for mitosis to occur, the nucleus has to move into the center of the cell. This happens during [[G1 phase]] of the cell cycle.

Initially, cytoplasmic strands form that penetrate the central vacuole and provide pathways for nuclear migration. Actin filaments along these cytoplasmic strands pull the nucleus into the center of the cell. These cytoplasmic strands fuse into a transverse sheet of cytoplasm along the plane of future cell division, forming the phragmosome. Phragmosome formation is only clearly visible in dividing plant cells that are highly vacuolated.

Just before mitosis, a dense band of microtubules appears around the phragmosome and the future division plane just below the plasma membrane. This preprophase band marks the equatorial plane of the future mitotic spindle as well as the future fusion sites for the new cell plate with the existing cell wall. It disappears as soon as the nuclear envelope breaks down and the mitotic spindle forms.

When mitosis is completed, the cell plate and new cell wall form starting from the center along the plane occupied by the phragmosome. The cell plate grows outwards until it fuses with the cell wall of the dividing cell at exactly the spots predicted by the preprophase band.

Preprophase Band

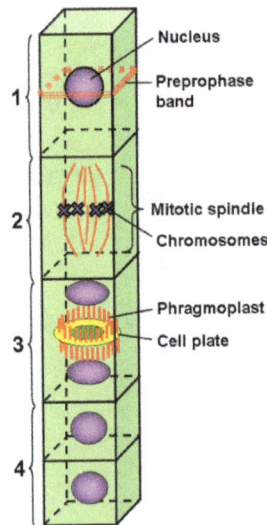

The preprophase band predicts the cell division plane: 1) Preprophase band formation during preprophase. 2) Metaphase spindle orients with the equator along the plane marked by preprophase band. 3) Phragmoplast and cell plate form along the plane marked by preprophase band. 4) The new cell wall of the daughter cells connects with the parent cell wall along the line of the former preprophase band location.

The preprophase band is a microtubule array found in plant cells that are about to undergo cell division and enter the preprophase stage of the plant cell cycle. Besides the phragmosome, it is the first microscopically visible sign that a plant cell is about to enter mitosis. The preprophase band was first observed and described by Jeremy Pickett-Heaps and Donald Northcote at Cambridge University in 1966.

Just before mitosis starts, the preprophase band forms as a dense band of microtubules around the phragmosome and the future division plane just below the plasma membrane. It encircles the nucleus at the equatorial plane of the future mitotic spindle when dividing cells enter the G2 phase of the cell cycle after DNA replication is complete. The preprophase band consists mainly of microtubules and microfilaments (actin) and is generally 2-3 μm wide. When stained with fluorescent markers, it can be seen as two bright spots close to the cell wall on either side of the nucleus.

Plant cells lack centrosomes as microtubule organizing centers. Instead, the microtubules of the mitotic spindle aggregate on the nuclear surface and are reoriented to form the spindle at the end of prophase. The preprophase band also functions in properly orienting the mitotic spindle, and contributes to efficient spindle formation during prometaphase.

The preprophase band disappears as soon as the nuclear envelope breaks down and the mitotic spindle forms, leaving behind an actin-depleted zone. However, its position marks the future fusion sites for the new cell plate with the existing cell wall during telophase. When mitosis is completed, the cell plate and new cell wall form starting from the center along the plane occupied by the phragmosome. The cell plate grows outwards until it fuses with the cell wall of the dividing cell at exactly the spots predicted by the position of the preprophase band.

Amyloplast

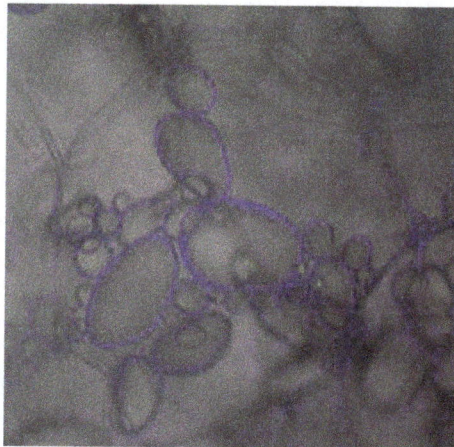

Amyloplasts in a potato cell

Amyloplasts are non-pigmented organelles found in some plant cells. They are responsible for the synthesis and storage of starch granules, through the polymerization of glucose. Amyloplasts also convert this starch back into sugar when the plant needs energy. Large numbers of amyloplasts can be found in fruit and in underground storage tissues of some plants, such as in potato tubers.

Amyloplasts are plastids, specifically leucoplasts. Plastids are a specialized class of cellular organelles that carry their own genome and are believed to be descendants of cyanobacteria (blue-green algae) which formed a symbiotic relationship with the eukaryotic cell.

Starch synthesis and storage also takes place in chloroplasts, a type of pigmented plastid involved in photosynthesis. Amyloplasts and chloroplasts are closely related, and amyloplasts can turn into chloroplasts; this is for instance observed when potato tubers are exposed to light and turn green.

Statoliths: Sensing Gravity

In the root cap (a tissue at the tip of the root) there is a special subset of cells, called statocytes. Inside the statocyte cells, some specialized amyloplasts are involved in the perception of gravity by the plant (gravitropism). These specialized amyloplasts—called statoliths—are denser than the cytoplasm and can sediment according to the gravity vector. The statoliths are enmeshed in a web of actin and it is thought that their sedimentation transmits the gravitropic signal by activating mechanosensitive channels. The gravitropic signal then leads to the reorientation of auxin efflux carriers and subsequent redistribution of auxin streams in the root cap and root as a whole. The changed relations in concentration of auxin leads to differential growth of the root tissues. Taken together, the root is then turning to follow the gravity stimuli. Statoliths are also found in the endodermic layer of the inflorescence stem. The redistribution of auxin causes the shoot to turn in a direction opposite that of the gravity stimuli.

Tobacco BY-2 Cells

Tobacco BY-2 cells is a cell line of plant cells, which was established from a callus induced on a seedling of *Nicotiana tabacum cv. BY-2* (cultivar Bright Yellow - 2 of the tobacco plant).

Overview

Tobacco BY-2 cells are nongreen, fast growing plant cells which can multiply their numbers up to 100-fold within one week in adequate culture medium and good culture conditions. This cultivar of tobacco is kept as a cell culture and more specifically as cell suspension culture (a specialized population of cells growing in liquid medium, they are raised by scientists in order to study a specific biological property of a plant cell). In cell suspension cultures, each of the cells is floating independently or at most only in short chains in a culture medium. Each of the cells has similar properties to the others. The model plant system is comparable to HeLa cells for human research. Because the organism is relatively simple and predictable it makes the study of biological processes easier, and can be an intermediate step towards understanding more complex organisms. They are used by plant physiologists and molecular biologists as a model organism.

They are used as model systems for higher plants because of their relatively high homogeneity and high growth rate, featuring still general behaviour of plant cell. The diversity of cell types within any part of a naturally grown plant *(in vivo)* makes it very difficult to investigate and understand some general biochemical phenomena of living plant cells. The transport of a solute in or out of the cell, for example, is difficult to study because the specialized cells in a multicellular organism behave differently. Cell suspension cultures such as tobacco BY-2 provide good model systems for these studies at the level of a single cell and its compartments because tobacco BY-2 cells behave very similarly to one another. The influence of neighbouring cells behavior is in the suspension is not as important as it would be in an intact plant. As a result any changes observed after a stimulus is applied can be statistically correlated and it could be decided if these changes are reactions to the stimulus or just merely coincidental. In this moment BY-2 cells are relatively well understood and often used in research. This model plant system is especially useful for studies of cell division, cytoskeletons, plant hormone signaling, intracellular trafficking, and organelle differentiation.

References

- Money, Nicholas P. (1995-12-31). "Turgor pressure and the mechanics of fungal penetration". Canadian Journal of Botany. 73 (S1): 96–102. ISSN 0008-4026. doi:10.1139/b95-231

- J., Kramer, Paul (2012-01-01). WATER RELATIONS OF PLANTS. Elsevier Science. ISBN 0124250408. OCLC 897023594

- Waggoner, Paul E.; Zelitch, Israel (1965-12-10). "Transpiration and the Stomata of Leaves". Science. 150 (3702): 1413–1420. ISSN 0036-8075. PMID 17782290. doi:10.1126/science.150.3702.1413

- Kinsman, R (January 1991). "Gas vesicle collapse by turgor pressure and its role in buoyancy regulation by Anabaena flos-aquae". Journal of General Microbiology: 1171–1178

- Fricker, M.; Willmer, C. (2012). Stomata. Springer Netherlands. p. 18. ISBN 978-94-011-0579-8. Retrieved 15 June 2016

- Hopkin, Michael (2007-07-26). "Carbon sinks threatened by increasing ozone". Nature. 448 (7152): 396–397. Bibcode:2007Natur.448..396H. PMID 17653153. doi:10.1038/448396b

- Beauzamy, Lena (May 2015). "Quantifying Hydrostatic Pressure in Plant Cells by Using Indentation with an Atomic Force Microscope". Biophysical Journal. 108: 2448–2456

- "Living Environment—Regents High school examination" (PDF). January 2011 Regents. NYSED. Retrieved 15 June 2013

- Schmerler Samuel, Wessel Gary (January 2011). "Polar Bodies - more a lack of understanding than a lack of respect". Mol Reprod Dev. 78 (1): 3–8. doi:10.1002/mrd.21266

- Petra Dietrich; Dale Sanders; Rainer Hedrich (October 2001). "The role of ion channels in light-dependent stomatal opening". Journal of Experimental Botany. 52 (363): 1959–1967. PMID 11559731. doi:10.1093/jexbot/52.363.1959

- "Fungal Cells Turgor Pressure: Theoretical Approach and Measurement (PDF Download Available)". ResearchGate. Retrieved 2017-04-27

- Pickett-Heaps JD, Northcote DH (1966). "Organization of microtubules and endoplasmic reticulum during mitosis and cytokinesis in wheat meristems". Journal of Cell Science. 1 (1): 109–120. PMID 5929804

- Charles E. Allen (July 1901). "On the Origin and Nature of the Middle Lamella". Botanical Gazette. 32 (1): 1–34. JSTOR 2464904. doi:10.1086/328131

- Eichorn, Susan, et al. Esau's Plant Anatomy: Meristems, Cells, and Tissues of the Plant Body: Their Structure, Function, and Development, 3rd Edition. 2006. ISBN 978-0-471-73843-5

- Lang, Ingeborg; Sassmann, Stefan; Schmidt, Brigitte; Komis, George. "Plasmolysis: Loss of Turgor and Beyond". Plants. Retrieved 10 March 2016

- Ambrose JC, Cyr RJ (2008). "Mitotic spindle organization by the preprophase band". Molecular Plant. 1 (6): 950–960. PMID 19825595. doi:10.1093/mp/ssn054

Permissions

Index

www.ingramcontent.com/pod-product-compliance
Lightning Source LLC
Chambersburg PA
CBHW062004190326
41458CB00009B/2968